Steps Towards a Unified Basis for Scientific Models and Methods

Inge S Helland

University of Oslo, Norway

Steps Towards a Unified Basis for Scientific Models and Methods

World Scientific

NEW JERSEY · LONDON · SINGAPORE · BEIJING · SHANGHAI · HONG KONG · TAIPEI · CHENNAI

Published by

World Scientific Publishing Co. Pte. Ltd.

5 Toh Tuck Link, Singapore 596224

USA office: 27 Warren Street, Suite 401-402, Hackensack, NJ 07601

UK office: 57 Shelton Street, Covent Garden, London WC2H 9HE

British Library Cataloguing-in-Publication Data

A catalogue record for this book is available from the British Library.

ISBN-13 978-981-4280-85-3

ISBN-10 981-4280-85-2

Printed in Singapore by B & Jo Enterprise Pte Ltd

Chapter 1

THE BASIC ELEMENTS

1.1 Introduction: Complementarity and Its Implications

Later in this book I will wind up with a rather extensive discussion of the basic principles of quantum mechanics. I want to argue that these principles, usually presented in a very formal way, can be linked in part to considerations related to common sense and in part to a relatively intuitive extension of ordinary statistical theory. One main point is: Every observation of nature, every interpretation of nature and every prediction of natural phenomena is done by some human being, and thus it is not unreasonable to expect that it should be possible to find some language for interpretation of nature — also for the basic physical nature — which has a relation to the everyday ordinary language used by humans. My aim is to show that such a basic formulation can be constructed. I will indeed show that essential parts of the ordinary laws of quantum mechanics may be derived from such a basis. The purpose of this programme is partly to work towards a unity of science, a unity where it may be a possibility for the various disciplines to learn from each other, partly to give an opportunity for a new discussion of the strange paradoxes connected to the ordinary quantum formalism, and more generally, to propose a conceptually simpler basis for the theory.

In fact, certain elements of quantum theory have already found its way into everyday language. I am then first thinking of the concept of *complementarity*, which is very central in every discussion of quantum theory. In the fascinating time at the first part of the previous century when ordinary quantum mechanics evolved from intuitive ideas towards more definite formulations, the precise meaning of this concept in the atomic setting was first developed by Niels Bohr. Already in these early days, Bohr gave lectures about the concept of complementarity to sociologists, to biologists, to people from medicine and to people from other natural sciences, emphasizing its precise meaning in the microworld, but also implicating that it could be useful in these other settings. At that time these implications were difficult to concretize in a concise way, but they were nevertheless of importance. Later, many physicists have been sceptical against a link towards the use of the word complementary outside quantum theory. The argument has been that the use of words should be precise, and often this has been taken to imply that every conceivable statement should be given a precise mathematical formulation, and that only such a formulation is acceptable. Later in this book I will argue for the view that intuitive ideas often should be the fundamental ones, and that the formal statements in some sense should be derivable from these. The task of deriving these connections will require precise mathematics, however.

Let us take a brief look at the concept of complementarity in the setting that it first arose. In quantum physics the position and velocity of a given particle may be regarded as complementary. A first formulation can be: For any given particle it is impossible to measure both position and velocity with infinite precision. A more precise way to formulate the (non-relativistic) quantum mechanical statement is given by Heisenberg's inequality, which can be stated as follows:

Theorem 1.1.1. *For a particle in some state let one have the choice between measuring a position component or a velocity component, both in the same direction. Let Δx be the standard deviation of position measurement and let Δv be the standard deviation of the velocity measurement. Then whatever the state should be and in whatever way the experiment is carried out, one must always have*

$$\Delta x \cdot \Delta v \geq \hbar/2m, \qquad (1.1)$$

where $\hbar$ is Planck's constant $(= 1.055 \cdot 10^{-34}\,Js)$ and m is the mass of the particle.

In formal quantum mechanics, Heisenberg's inequality is a result of the non-commutivy of the position and the momentum-operator. A close analogy in theoretical statistics is given by the Cramér-Rao inequality. I will come back to this in Chapter 7 and will indicate that it can be more than an analogy.

A qualitative verbal consequence of Heisenberg's inequality may be: It is impossible to measure both position and velocity with infinite precision, but the limitation posed for a joint precise measurement will be lower if the mass is increased. I will claim that such qualitative statements in fact can be very useful, and I will also take the step of generalizing by analogy to other situations. In the position/ velocity case and in many other physical cases the qualitative statement of the case indicated above can be made more precise by formulating inequalities of the type given by (1.1).

A new, deeper and more precise derivation of Heisenberg's inequality has recently been put forward by Busch and coworkers [39].

The mathematical quantum mechanical literature of the last 80 years has also worked for precision, but then in a different direction: In the current formulation all observable quantities are associated with operators on a given abstract space, and general inequalities of the type (1.1) are constructed using the commutators of these operators. This mathematical theory is very general, and it is the starting point of much theoretical physics and much pure mathematics. I will aim at linking my own approach to this formal world in later chapters, but my emphasis in this book will at least to some extent be completely different: I regard physics as an empirical science, and I feel that one should try to base physics as a science on intuitive ideas, ideas that can be linked to other empirical sciences. In particular, the process of measurements in physics should have a close relationship to the way measurements are made in other sciences.

One can even argue, in the spirit of the above discussion, that simplicity in terms of an intuitive understanding of physics together with the rest of science on the one hand and precision and generality in terms of formal mathematics on the other hand in part may be regarded as complementary intentions. To draw the analogy with (1.1) even further, one can say that the aim of the present book is to contribute to increasing the 'mass' of this particular complementarity relation. At least, my intention is to initiate a development in such a direction, with the hope that this will inspire other scientists to work along the same lines.

Measurements in science are nearly always uncertain, and they are nearly always made to estimate some theoretical quantity. Good estimators in statistics are usually found using some optimization criterion: Least squares, maximum likelihood, highest information and so on. This is discussed in many statistics textbooks, always in a situation where one single experiment is treated, and always where a single experiment is defined in the common statistical sense: All measured quantities are defined as random variables on a single probability space. Later, we will discuss this limitation, introducing the concepts of counterfactuals and focusing.

In practice any scientist in a given complicated enough situation will have the choice between several experiments or variants of the same experiment. Even in the case of a single measurement series, 'complementary' considerations may be of importance, say between buying an accurate expensive measurement apparatus or investigating many units.

Similarly, when one of several potential experiments are being considered using limited resources, this may result in a situation where different estimators result from optimizing complementary criteria in different, mutually excluding experiments.

Thus the concept of complementary values may seem to be important when resources are limited. And in some sense or another, resources are nearly always limited. Thus the concept of complementary values, admittedly imprecise at this stage, seems to generalize to a large number of situations.

Let us now make another quantum leap concerning these rather informal application of the concept of complementarity. We claim that the basic mode of understanding found within the world's different scientific communities to some extent are complementary. The communication between chemometricians, statisticians, physicists and mathematicians, say, as it is today, could have been much better. Different historical developments have resulted in different traditions and in different scientific languages in the various disciplines. The result has been scientific communities with partly complementary modes of thinking. To be precise, however, the world 'complementary' is used here in a slightly different sense than its use in physics.

Increasing the relevant 'mass' here must of course necessarily again be a very long term project. In my opinion, the most fruitful approach to this end seems to be to try to take as points of departure simple concepts that everybody can relate to, while in the same time trying to make the concepts precise within a modelling framework.

Aiming towards such a goal will require quite a lot of care in the formulation used. At least at this stage one may at some times err in the direction of using too simple formulations rather than too involved and precise formulations. For some readers this may mean that I at certain places may tend to trivialize concepts and results that they see — quite rightly — as the result of a sophisticated scientific development. On the other hand, other readers may find that my arguments at places are too mathematical. For both these kinds of readers I sincerely apologize.

Finally, it must be said that any book of course is written by an author or by authors with specific backgrounds. This particular book is written by a statistician, but one that also has tried to keep his eyes open also to other scientific traditions. Regardless of personal background, however, it is my conviction that the statistical language can be made suitable for finding a basis for a large number of sciences regarding the way inference is made from data. This conviction is a result of the simple empirical fact that statistical methods have proven to be useful in many of these sciences, and then a naive deduction that this must imply that the process of relating observations to the state of nature must contain some common basic elements across sciences. It is clear, however, that to reach a complex theory like quantum mechanics from such a point of departure, the statistical language must be extended. A beginning of such an extension will be given here, and this will be continued in the Chapters 4, 5 and 6.

Most empirical sciences, for instance biology and medicine, use to a large extent rather simple statistical methods. Hence most of my examples, at least initially, will relate to situations where such methods are used. This should not preclude, however, that this discussion also should have implications to situations where more sophisticated statistical methods are needed. However, these latter situations are often so dominated by complicated mathematics, modelling problems, data organization and/ or computational issues that the methodological themes discussed here may tend to be overlooked. This is also a good reason to stick to simple examples.

Before turning to a review of statistics, however, I will go through some basic concepts needed in the sequel. The first task will be to choose a framework for the language used, a choice that must be made before defining the necessary words.

1.2 Conceptually Defined Variables

To describe elements of nature in the way we really think they are, we need a language. This language will be different for different subject matters and for different investigations, but very often the discussion ends up by focusing on certain conceptually defined variables, variables which in some way or other are to be confronted with empirical observations. Such conceptually defined variables are well known in statistics, as parameters of statistical models, but are less used in physics. Later I will use them repeatedly in my approach to quantum mechanics. Two simple examples may be the velocity and the position of a particle in the way these concepts are used in Heisenberg's inequality. To begin with, these are conceptually defined variables; they turn into observable quantities only when experimental settings are defined and executed.

In many statistical investigations the conceptually defined variables will be parameters of some model. These are usually written as Greek letters ξ, θ, μ, σ,.... There is of course nothing sacred about the Greek letters, but this convention is very useful when one wants to distinguish between the state variables of a system — written by Greek letters -, and observations - given by some Latin letters y, z,....

The same distinction will be aimed at throughout this book, but the scope will be extended. I find it appropriate to introduce a new general term: conceptual variable or c-variable, which is an extension of the statistical parameter term. I will discuss this in detail later; for statisticians I will just mention a few examples here: In a statistical hypothesis testing setting one can define a parametric variable which is 1 when the hypothesis should be rejected and 0 otherwise, and name this a c-variable. As extensions of this one can introduce c-variables in multiple testing situations, for different choices of sets of orthogonal contrast in analysis of variance, for different choices of conditioning and so on. All this will be further discussed later.

In quantum physics it will turn out to be useful to use the term c-variable for conceptual values of position, velocity and spin component, and for the cartesian product of such variables. At several places in the book the c-variables — Greek letters — will be associated with conceptual quantities, as opposed to concrete, observed quantities. This way of putting the distinction is in fact very useful in many empirical investigations. I am aware that this way of thinking is unfamiliar to physicists. They are used to thinking of variables as operators and concrete values as eigenvalues of these operators. I will come back to this formal apparatus in Chapter 5, and then derive it from my approach.

Definition 1.2.1. *A c-variable is some conceptual quantity connected to a system or a population. The purpose of an empirical investigation is often to arrive at some statement concerning certain c-variables or on some parts of a given c-variable.*

Fig. 1.1   c-variable.

In many statistical textbooks the parameters are always connected to some infinite population. Concerning c-variables I will not make such a connection in this book, and this point is essential. As a typical example, let the real weight of some object be μ, and let us make independent measurements $y_1, y_2, \ldots, y_n$ of it. Then one often estimates the weight μ by $\hat{\mu} = \bar{y} = n^{-1} \sum y_i$. It is of course important in this connection to distinguish between real weight and estimated weight. Of course one can talk about an infinite population of measurements here, but from a physical point of view, the main issue is that we consider inference from a finite amount of data.

There are many variations around this theme. For instance, assume that in some population, the expected weight loss of a person is μ_1 under diet 1 and μ_2 under diet 2. Select n_1 persons that are given diet 1, and n_2 persons that are given diet 2, and let their measured weight losses after being exposed to the diets be $y_{11}, \ldots, y_{1n_1}$, respectively $y_{21}, \ldots, y_{2n_2}$. Then, on the basis of $\bar{y}_1. - \bar{y}_2.$, where $\bar{y}_i. = n_i^{-1} \sum_j y_{ij}$, and some probability model for the observations, a common task is to say something about the sign and size of $\mu_1 - \mu_2$.

An example more in the spirit of the previous example is when you have two related objects with real weights μ_1 and μ_2, respectively, and make repeated independent measurements on each object. It is important to realize that we in this book also connect c-variables to finite objects.

In physics we sometimes do not have repeated measurements. Also in that case we must distinguish between the c-variable θ and the corresponding observation $y = \hat{\theta}$, for instance the velocity of a particle and the observed velocity under some experimental setting.

In principle, every observation requires some measurement apparatus. In some cases this measurement apparatus is perfect for practical purposes, so we can disregard measurement error. Then θ and $y = \hat{\theta}$ are numerically equal. But even in such an instance of *perfect measurement*, it may in some cases be important to distinguish between c-variable and observation, a point I will come back to in Chapter 2 and then later in connection to the choices of experiments in quantum mechanics. But one conceptual clarification related to this fact may be discussed already now.

1.3 Inaccessible c-Variables

The model description above demands in principle that every c-variable should be estimable from the available data. For models involving only one or a few individuals, this is typically not realistic at all. Here are some examples. Some standard statistical concepts are used here in a nonstandard setting; for a definition of these concepts, see sections 1.8 and 1.9 below. Note that these examples are meant as statistical examples; the link to quantum mechanics will come much later.

Example 1.3.1. We want to measure some quantity with an apparatus which is so fragile that it is destroyed after a single measurement. We may model the measured values to have an expectation μ and a standard deviation σ, perhaps even a normal distribution with these parameters. A single measurement gives an estimate of μ. The standard deviation may be thought to be possible to estimate by dismantling the apparatus, again destroying it. In a certain sense we might say that in this example, the parameters μ and σ are complementary: Only one of these can be estimated.

Nevertheless, define the two-dimensional c-variable (μ, σ) as ϕ. As an abstract vector, this can in a sense be meaningfully defined, but this vector can never be given a value, in the sense that there is no empirical mechanism through which we can approach the process of finding a value for the vector. Such cases in fact exist in many different connections, and an observation like the one above will turn out to be very important for us later.

Example 1.3.2. Assume that two questions are to be asked to a single individual at some given moment, and that we know the answer will depend on the order in which the questions are posed. Let the conceptual variable or parameter (θ^1, θ^2) be the expected answer when the questions are posed in one order, and (θ^3, θ^4) when the questions are posed in the opposite order.

Then the c-variable $\phi = (\theta^1, \theta^2, \theta^3, \theta^4)$ cannot be estimated directly from one individual, that is, it cannot be given a value.

Example 1.3.3. Consider the location and scale parameters μ, σ in a case where only measurement on a single individual is possible. Say, let μ be the expected measurement of some position, and let σ be the standard deviation of that measurement. This is somewhat similar to Example 1.3.1 above, that is, if we really insist that the firm context is such that only one measurement on the individual is possible, then it is impossible to give a value to the vector (μ, σ).

Example 1.3.4. A given patient at a given time has (expected) recovery time θ^1 if treatment 1 is used and θ^2 if treatment 2 is used. The term expected here (and in the examples above) must be interpreted in some loose sense, not necessarily with respect to a well-defined probability model. Rather, we can imagine θ^1 to be the recovery time that an extremely experienced medical doctor will expect given everything that is known about treatment 1 and given everything relevant that is known about the patient. Similarly for θ^2.

Like all parameters, θ^1 and θ^2 can be estimated from experiments, but it is impossible to estimate both parameters on the same patient at the same time. Nevertheless, consider the c-variable $\phi = (\theta^1, \theta^2)$. In this case we can imagine ϕ to have a value if this is assessed by an extremely experienced and reliable medical doctor. But in a context where we doubt the judgement of that doctor and want empirical evidence, ϕ can never be given a value. In any case, ϕ must be considered to be *inaccessible*: It can never be estimated from data.

In all the examples above the situation can be amended through investigating several individuals, but this assumes that the parameters are identical for the different individuals, a simplification in many cases. Our point with these examples is not that they cannot be modified to avoid inaccessibility of the c-variables. In fact they always can. Our point is more that such examples exist. Later, such c-variables wil constitute one of the presuppositions for deriving quantum mechanics from a non-formal basis. Note that the ordinary statistical paradigm in the simplest case assumes an infinite population of individuals with the same parameters. This will not be assumed in the present book.

Many more realistic, moderately complicated, examples exist, for instance, consider behaviourial parameters of a rat taken together with parameters of the brain structure which can only be measured if the rat is killed. Even more complicated examples can be imagined in public debates where different discussants take incompatible concepts as points of departures, or in novels where items are elucidated from several different angles.

Definition 1.3.1. *When considering these cases where ϕ cannot be estimated from any experiment on the given individuals, we may call $\Phi = \{\phi\}$ an inaccessible c-variable space rather than a parameter space, and ϕ is called an inaccessible c-variable. Any function of ϕ that cannot be estimated from experiment, will be called inaccessible.*

Accessible c-variables may be of one of two kinds:

1. They may be estimable parameters in some statistical model; or

2. They may correspond to a limiting case where measurement uncertainty may be disregarded. In this latter case the c-variable takes values in some given set, either continuous or discrete.

I will insist that modelling through inaccessible c-variables can be enlightening, in the cases mentioned above as in other cases. In particular this may be useful if one models cases where one has the choice between several measurements, as one usually will have in quantum mechanics. As will be discussed below, by choosing a particular experiment in a given setting, what one can hope for, is only to be able to estimate a specific (accessible) part of the inaccessible c-variable.

The concept of complementarity is well known in quantum physics, and the concept of complementary c-variables is central to the way I later intend to introduce quantum theory. Here is an example already mentioned from physics: In a model of a particle we can imagine that it has a theoretical position $\theta^1 = \xi$ and a theoretical momentum $\theta^2 = \pi$, but there is an absolute limit to how accurate these parameters can be determined jointly. In fact, these quantities must always be considered as theoretical parameters, jointly as c-variables. By the well known Kochen-Specker theorem one cannot assign pre-existing values to all properties of a physical system. Similarly, we can let λ^1 and λ^2 be spin components of a particle in two different selected directions, but only co-insisting as c-variables.

Thus, in this paper I will allow c-variables like the vectors $\phi = (\mu, \sigma)$, $\phi = (\theta^1, \theta^2)$ or $\phi = (\lambda^1, \lambda^2)$ in the examples above that are not identifiable in any model and consequently can not be estimated by performing any given measurements on the units of a population. The way to make inference possible anyway, is to use model reduction, i.e., focus on subvariables (parameters) and a corresponding choice of experiment. This choice of subvariables will be important in the discussion which follows.

All this will be done together with a relevant choice of symmetry group on the inaccessible c-variable space. These are the main points in what I claim is a way to understand quantum theory from a statistical point of view. Some additional technical assumptions will be formulated later. Of particular importance will be an assumed symmetry relationship between the experiments corresponding to incompatible c-variables as a basis for quantum theory. But this I will come back to in Chapters 4 and 5.

It is of interest that several other authors also have aimed at a more intuitive basis for the quantum formalism. Some of the most interesting of these endeavours have taken information theory as a point of departure. This was a development which started with Wotters [209]. One of the most recent papers here is Goyal [99], where further references can be found. Another approach is based on decision theory (Finkelstein [81]).

1.4   On Decisions from a Statistical Point of View

Decision theory is central to the theory of statistical inference, and is discussed in many textbooks. A natural point of departure here is the book on Bayesian theory by Bernardo and Smith [26]. Here a decision problem is defined as consisting of 3 elements:

  - a set of available actions, one of which is to be selected;
  - for each action, a set of uncertain events, describing the uncertain outcomes of taking a given action;
  - for each such event, a certain consequence.

A typical example is the estimation of a parameter. The set of actions is the set of values of the parameter, and an estimator is a function from the data to this set. Given the data, the decision, the estimate is the value of this estimator at these data. Each possible estimator has its uncertainty. Often the estimate is found by minimizing a loss function. A simple consequence is then the value of the loss function at a given value of the estimator.

Later in this book we will need a more complicated structure for our decisions, a structure which is not too far away from everyday experience: Before we are ready to choose an action, we must focus upon one of a limited set of possibilities. In fact, this can be looked upon as a two-step procedure, where the first step has the above structure, namely, the choice of focus is looked upon as a decision problem of its own merits.

Specifically, consider a situation where we have an inaccessible c-variable $\phi = (\lambda^a; a \in \mathcal{A})$ containing parameters λ^a, each corresponding to a certain experiment, where a runs through an index set $\mathcal{A}$. An example from quantum mechanics as we will see it later, may be where λ^1 and λ^2 are the spin components of a particle in two different directions. Then $\phi = (\lambda^1, \lambda^2)$ must be considered as a c-variable, and one must make a choice, focusing upon λ^1 or λ^2 before making a measurement.

General procedure. *The decision process will in general run through the following steps:*

  1. Choose a focus a, that is, a parameter λ^a and the corresponding experiment;

  2. Perform the selected experiment and obtain data x^a;

  3. Within the selected experiment, and based on the data obtained, one can make certain inference decisions, say estimation of a set of parameters. This will lead to uncertain events in the range space of the estimator, and each decision implies consequences which can be anticipated, and possibly be taken into account when the decision shall be made.

There are many variations of this general procedure. In some cases, the estimation under 3) involves all parameters of a statistical model, and λ^a is a fixed dimension parameter which is estimated independently. We will meet such a case in Chapter 8. In other cases, the choice of the parameter λ^a is subjective, like in the particle spin case. In each single case, however, a sequence of decisions is involved. It is important that this sequence does not bias the final inference step. A necessary condition for this is the following:

Unprevision condition. *The decision taken at some given step should not utilize any information about a part of the c-variable or on the data which is used in decisions taken in later steps.*

Example 1.4.1. In any given classical experiment, that is, a class of probability distributions of some observations indexed by some parameter, it has been recommended by Fisher to make any estimation by first taking the conditional expectation on any ancillary statistic. *An ancillary statistic is a function of the observation whose distribution does not depend upon the parameter.* This conditioning upon an ancillary may be looked upon as a two-step procedure, where the first step is to choose the value of the ancillary statistics. In other cases we also have the choice between several possible ancillaries. By its very nature, the value of an ancillary statistic contains no information about the parametric function that one wants to make inference upon, so this is in agreement with the unprevision condition. Also, according to the unprevision condition, any criterion for choosing between ancillaries (e.g., Cox [49]) must have a similar property, i.e., be independent of the parameter.

Example 1.4.2. A c-variable ϕ is a vector: $\phi = (\lambda^1, \lambda^2, ..., \lambda^A)$. For each a there is a maximal experiment with parameter λ^a. Choosing some fixed a independently of everything else will lead to honouring of the unprevision condition. This example can have a classical or quantum mechanical interpretation.

Example 1.4.3. As a part of a larger medical experiment, A individuals have been on a certain diet for some time, and by taking samples at the end of that period some response like the change in blood cholesterol level is measured. For the individual a ($a = 1, 2, ..., A$), the measured response is x^a, which we suppose has expectation μ^a and a known measurement variance σ^2. Measurements on different individuals are independent.

Assume that we are interested in estimating the mean expected value

$$\mu = \frac{1}{A}(\mu^1 + \mu^2 + ... + \mu^A).$$

Then we can imagine two different scenarios:

1. We select randomly some $M < A$ individuals, and are just given the numbers $x^1, x^2, ..., x^M$, but do not know anything more about the individuals associated with these numbers. Then we might well imagine that there is a decision procedure behind the choice of these numbers, but this decision procedure does not rely on any information of relevance to the estimation of μ. The unbiased estimate is simply

$$\hat{\mu} = \frac{1}{M}(x^1 + x^2 + ... + x^M).$$

2. Again we select $M < A$ individuals, but now we know which individuals that are selected; these are pre-determined. Then it is impossible to find an unbiased estimate of μ from the available data. The simplest case is when $M = 1$. Then we know that some individual, say with expectation μ^1 has been selected. By the conditionality principle, which will be further discussed in Chapter 2, every inference should be conditional upon this choice. Thus in this case we just have data x^1, and can only make inference on the simple parameter μ^1. Nothing can be known about the average parameter μ, and a similar conclusion is true when $1 < M < A$. (In the case $M = A$ we of course have full information.) A general way to explain the lack of information in case 2, is that the prevision condition is violated.

A perhaps surprising conclusion from this example is that there exist cases where one can make better statistical inference when one knows less.

For statisticians: Further examples of relevance to the prevision condition are given by randomization in design of experiment situations, also by the selection of contrasts in an analysis of variance. These examples will be discussed in Chapter 7.

1.5 Contexts for Experiments

Any experiment is done in a context, that is, for some given experimental units, any preparation done on these units, any assumptions explicitly made on the units and on the statistical model, preferably verified before the experiment or justified by external arguments, and finally any environment chosen for the whole experiment. The results found in earlier experiments may also contribute to the context.

A part of the context may even be the routine built up by the experimentalists, or their prejudices towards the expected result. To compensate for the latter, in medical experiments one is recommended to have a double-blind setup, i.e., one where neither the patient nor the experimentalist knows which treatment is used on the patient. For practical reasons it is not always possible to achieve this ideal, however.

Thus, according to the setting, one may be able to use parameters which are independent of the context, or is forced to let at least some of them depend upon the context. This latter situation will in most cases lead to less informative estimation. One may in given cases be able to model this dependence, however.

I will come back to the concept of context on some occasions later. A part of the context may consist of assuming that certain parameters in the statistical model have some fixed values. This may be the case when the initial model has many parameters, and it is desirable to do some model reduction. More generally, the context may have implications for the statistical model chosen or for the way that one chooses to perform an experiment.

Considerations of this kind are relevant to physical experiments, too. The concept of context is discussed at length in Khrennikov [135, 136], where this concept is connected to a mathematical formalism. No such formalism will be used at the outset in this book, but I will later come back to formalisms on the basis of the conceptual discussion started here.

As already indicated in the introduction; see the discussion around Theorem 1.1.1, certain pairs of experiments may be *incompatible*:

Definition 1.5.1. *Two experiments are called incompatible if only one of these experiment can be performed in a given context.*

Niels Bohr used the concept of complementarity in a sense closely related to this. However, he considered experiments as constituting a context, so for physicists it may be confusing to talk about experiments *in* a context, a standard statistical notion. Many physicists have followed Bohr's use of the word complementarity even though it is somewhat problematic: The same word is used in a different meaning among other places in psychology (where it is the idea that people seek others whose characteristics are different from and complements their own) and in colour theory (where it is standard to talk about complementary colours). Some theoretical physicists, among them L. Accardi (oral communication), argue that the word complementarity should only be used for potential experiments that are maximally incompatible in a precisely defined sense: For two discrete parameters, in a state defined by the values of one of them, the posterior distribution of the other should be uniform over its values. By taking limits, a similar notion can be defined for continuous parameters, even parameters like expected position or momentum, taking values on the whole line.

1.6 Experiments and Selected Parameters

Every experiment begins with a question, that is, a precise formulation of the issue that we intend to investigate during the experiment.

Assume that the system under investigation can be described by a c-variable ϕ. The experiment $\mathcal{E}^a$, or simply the experiment a, may then be directed towards a specific parameter $\theta^a = \theta^a(\phi)$. The experimental question from this can simply be: What can we say from observations about the value of θ^a? It is important to be aware of that, depending upon the situation, the parameter θ^a may be multi-dimensional, so that several scalar parameters may be under investigation in the same experiment. In the simple first examples below we let θ^a be one-dimensional, however.

Look first at Example 1.3.1 concerning the fragile apparatus: What can we say from a chosen observation about the expectation μ? Or, what can we say about the standard deviation σ?

As another example (cf. Example 1.3.4), given one patient, what can one say about the expected recovery time θ^1 when treatment 1 is used? Or about the expected recovery time θ^2 when treatment 2 is used?

In a practical situation, one is often interested in a population of potential patients, and then the situation is much easier from a statistical point of view. A randomized experiment can be set up, see below, so that n_1 selected patients can be allocated to treatment 1 with expected recovery time μ_1 and n_2 patients to treatment 2 with expected recovery time μ_2, and inference can be made on μ_1, on μ_2 and finally on $\tau = \mu_1 - \mu_2$, which may be of interest. The full parameter of the experiment may be $\theta = (\mu_1, \mu_2)$, or it may contain more parameters.

The quantum-mechanical situation is usually concerned with a single unit or a few units, so our modelling must be geared towards that. But also in this case we assume a c-variable ϕ and then corresponding to the experiment $\mathcal{E}^a$ a selected parameter θ^a which again can be a scalar or a vector, i.e., in the latter case consist of several subparameters. A simple example is: θ^1=position and θ^2=momentum.

It should be clear from the examples of this chapter that it is seldom natural to assume that every function of the c-variable can be taken as a selected parameter θ^a connected to a concrete experiment.

Once the experiment/ selected parameter is chosen, the experiment can be performed and give observations y. These observations can at the outset be of different nature, but they can always be thought of as positions of pointers of some apparata, i.e., having some Euclidean coordinates.

After the experiment is performed, statistical inference can be used to say something about the value of the parameters of the experiment. In some sense: Before the experiment we know nothing about the parameter; afterwards we can have quite detailed knowledge. Admittedly: this can only be said in classical, frequentist statistics; in Bayesian statistics the prior may be said to specify a pre-experimental knowledge. Anyway, in our setting we feel that the situation can be related to the following citation by Bohr (cited from [174]): 'It is wrong to think that the task of physics is to find out how nature is. Physics concerns what we can say about nature.'

1.7   Hidden Variables and c-Variables

During the history of quantum mechanics so-called hidden variable theories have been introduced at various stages. Different authors have proposed theories where the usual formalism of quantum theory is supplemented in different ways by variables that are hidden in such a way that their values can not be revealed through experiments, or can only be revealed through certain specific experiments. Such hidden variables are crucial for the interpretation of quantum mechanics given by Bohm [29], for instance. The hidden variable theories have been rejected by many physicists since von Neumann [162] showed that they under certain circumstances lead to contradictions, but have become more popular again after Bell [22] pointed out that the assumptions made by von Neumann were not always reasonable.

Fig. 1.2   Paint.

Our c-variables ϕ are related to the hidden variables introduced by various physicists, but they may also be said to be different. The main common property relates to the fact that certain functions of the c-variables, the statistical parameters, are unknown before the relevant experiment is performed, and can be given a fairly accurate value afterwards. Later we will introduce perfect experiments, where every measurement error can be ignored, so that the value of the relevant parameter then is exactly known.

A main difference between c-variables and hidden variables relates to the fact that a c-variable, in the way we define it, does not necessarily have a value at all. Several examples of this were given in Section 1.3. To begin with, as discussed in Section 1.3, these c-variables may be viewed as defined by some formulation in ordinary language, but they can not always be given some specific value. A function of the c-variable may or may not have the same property.

A c-variable θ is defined through some verbal formulation, perhaps connected to some specific context. If this context is changed or if it is not realized, the definition of θ may cease to have a meaning.

Example 1.7.1. Define the c-variable θ as the expected drying time for the paint used when I repaint my house. This θ may not have a value if I choose definitively not to repaint my house at all. This may be regarded as a rather trivial example, but it illustrates our main point here: There does exist conceptually defined variables that do not have a value. Of course, a given brand of paint will have a definite expected drying time. But a person which has chosen not to repaint his house, may also not have chosen his brand.

In general, if θ is some c-variable, connected as a parameter to an experiment, it may or may not have a value if it turns out to be not to be a well-defined way to perform that experiment. It may have a value if this can be connected to other considerations, like if the paint is connected to some brand. But what is important is that a c-variable is only guaranteed to have a value if it can be connected to an experiment which is actually possible to perform.

As an extreme example of a verbal formulation which turns out not to make sense, consider the following:

Example 1.7.2. Let θ be the truth value of the sentence: 'What I say now is a lie'. This is of course a statement which does not have a definite truth value.

Then θ will never have a defined value. If the sentence is true, it is a lie, and if it is a lie, it is true. What I want to illustrate by this is: Sentences can be formulated by ordinary language, seemingly precise — in fact such sentences can be formulated in the language of formal logic also —, which can never be given a meaningful interpretation. Paradoxes of this kind have deep implications, and they have had a central place in formal logic since they were introduced there by Russel, Grelling and others. In our setting they can be chosen to represent the ultimate sense in which formally well-defined sentences can be taken not to have an operational meaning. These considerations can in our opinion also be taken to have some implications to the assertion that c-variables do not necessarily have values, since such c-variables always are defined by statements using ordinary language.

As a possible way to carry the theory developed in this book further, one could conjecture that it should be possible to define the concept of c-variable in a mathematically precise way using formal logic, a definition which would include the possibility that the parameter does not necessarily have a value, and would take into account the influence of the context on the state. Such a definition would then play the same rôle as Kolmogorov's precise mathematical definition of a random variable (see Section 1.8 below), and would place the development of both statistical theory and quantum theory indicated here on a firm mathematical foundation. For our purpose here, however, the intuitive content of the concept of c-variable is more important than a formal definition (in fact, the same is true for the concept of random variable), so we leave this task to future research.

In quantum theory the discussion above can in our opinion be linked to the Kochen-Specker theorem; for a precise statement of this see for instance [130]. The theorem rests upon technical concepts like observable, state function and Hilbert space, which we will define in Chapter 5, but which at this stage can be found in any book on quantum mechanics. Briefly, a system is defined by a Hilbert space $\mathbf{H}$, a state v is a vector of that space, and an observable A is an operator on $\mathbf{H}$. The observable A is supposed to be connected to some physical variable, but at this point it is not quite clear what this means in the general case when v is not an eigenvector of A.

As an attempt to make this clearer, one can tentatively introduce a value function $V_v(A)$ which is interpreted as the value of the physical quantity A when the system is in the state given by v. A natural requirement then is that

$$\mathcal{F}(V_v(A)) = V_v(\mathcal{F}(A))$$

Fig. 1.3   Chameleon.

for any realvalued function $\mathcal{F}$ on the real line. This implies a number of other plausible assumptions on any value function; see [130] for details. The Kochen-Specker theorem, which has a fairly long technical proof, then says that no such value function exists if the dimension of the Hilbert space is larger than 2.

Here is one last macroscopic example of a c-variable which does not have a value: Suppose that we have captured a chameleon, and define θ to be the colour of that chameleon. Then θ is only well-defined when we at the same time specify the base on which the chameleon stays. As is well known, a chameleon changes colour according to the place on which it sits. On a leaf it is green, on a branch of the tree it turns brown. This metaphor has been used extensively by Accardi [1–3] in his discussion of the interpretation of quantum mechanics. Of course this example can be bypassed by noting that the system here consists of the chameleon plus the base, but in any case it is a fact that the conceptually defined variable: the colour of the chameleon, does not have a value.

Our starting point is in agreement with this metaphor. But our foundation is different; most importantly, it is derived from common sense and a relatively wide range of examples, not from the formalism of ordinary quantum theory. One purpose of this book is to argue that the natural logic goes from simple concepts to more formal concepts, not in the opposite direction.

1.8   Causality, Counterfactuals

Look again at the example where θ is the expected drying time of the paint used if I repaint my house. If I definitely decide not to repaint the house, then this θ may be defined as a counterfactual quantity. In general, a counterfactual quantity is a quantity defined in a situation B, and is derived from an implication from situation A to the situation B together with a definite knowledge that A does not happen. As the example shows, such quantities may or may not have well-defined values, dependent upon the circumstances. Specifically, if I in the example already have decided always to use a particular brand of paint with a known expected drying time, then θ will have a meaningful value, otherwise it will probably not have a value.

In Section 1.3 I introduced the concept of an inaccessible c-variable, and gave several examples of it. The formal definition of such a c-variable ϕ was given in Definition 1.3.1, and it was also tied to the concept of complementarity. Then in Section 1.5 I discussed the context of an experiment, and indicated that the inaccessibility of a given c-variable may be due to the limitation caused by the context to the choice of experiment.

In my view this choice of experiment can also be related to the literature on causal inference, in particular to the concept of counterfactuals, which has a central place there. A counterfactual question is a question of the form: 'What would the result have been if ...?'. A counterfactual variable, in the way this concept is used in the literature, is a hypothetical variable giving the result of performing an experiment under some specific condition a, when this condition a is known not to hold. A typical example is when several treatments can be allocated to some given experimental unit at some fixed time, and then in reality only one of these treatments can be chosen.

The use of such a concept in the statistical literature goes back to Neyman [164], and has in the last decades been discussed by among others Rubin [180], Robins [178, 179], Pearl [171] and Gill and Robins [96]. On the other hand Dawid [53] is sceptical to a too extensive use of counterfactuals; he argues that causal inference should be performed without counterfactuals. The discussion of the latter paper shows some of the positions taken by several prominent scientists on this issue.

In our setting, we choose and perform one experiment a, and then any other experiment b imagined at the same time must be regarded as an experiment involving counterfactual variables. However, instead of introducing counterfactual variables, we use a conceptually defined variable or c-variable $\phi = (\lambda^a, \lambda^b)$ for the corresponding concept. A c-variable is a hypothetical entities that usually cannot be observed directly. Nevertheless they may be useful in our mental modelling of phenomena and in our discussion of them. In the last decades, such mental models in causal inference have been developed to great sophistication, among other things by using various graphical tools (Lauritzen [143]; Pearl [171]). In the present book we will limit mental models to scalar and vector parameters, some counterfactual, leading to our c-variables, but this model concept can in principle be generalized.

When it is decided to perform one particular experiment $a \in \mathcal{A}$, the λ^a becomes the parameter of this specific experiment, an experiment which then also may include a technical or experimental error. In any case the experiment will give an estimate $\hat{\lambda}^a$. If the technical error can be neglected, we have a *perfect* experiment, implying $\hat{\lambda}^a = \lambda^a$.

We are here at a crucial point for understanding the whole theory of this and the next chapter, namely the transition from the unobserved c-variable to the observed variable. Let us again look at a single patient at some given time which can be given two different treatments. Define λ^a as the expected survival time of this patient under treatment a. Then make a choice of treatment, say $a = 1$. Ultimately, we then observe a survival time t^1 for this patient. There is no technical error involved here, so we might say that we then have $\lambda^1 = \hat{\lambda}^1 = t^1$. *And this is in fact true.* Per definition, λ^1 is connected to the single patient, the definite treatment time and a definite choice of treatment. So even though λ^1 is defined at the outset as an unknown c-variable, its definition is such that, once the experiment is carried out, the c-variable must by definition take the value t^1.

This simple, but crucial phenomenon, which is related to how a concept can be defined in a given situation, is in my view of quantum mechanics closely connected to what physicists call 'the collapse of the wave packet' when an observation is undertaken.

1.9 Probability Theory

It is a strange historical coincidence that a definite axiomatic treatment of ordinary probability theory (Kolmogorov [137]) did not appear until 1933, the year after von Neumann [162] had axiomized quantum theory, a rather exotic looking theory in comparison. By now, probability theory is of widespread use, not only as a foundation of statistical theory, but in a multitude of other fields, including both foundational aspects and applied aspects of quantum mechanics. Some theoreticians argue that other axiomations, like that of von Mise or of Savage should be used in the foundation of quantum theory, but that will not be found necessary in this book. In fact, when it comes to it, the derivations from these various axiom systems are closely related. Kolmogorov's basic axioms for probability theory are given in Appendix A.1.1, where also the concept of a σ-algebra $\mathcal{F}$ is defined.

A further important concept is that of a conditional probability. In fact, since all probabilities are computed in a context as defined in Section 1.5 and therefore are calculated under some assumption, it can be argued that all probabilities used in practice are conditioned in some way. The formal definition runs as follows:

Definition 1.9.1 *Assume that $A, B \in \mathcal{F}$ and $P(B) > 0$. Then the conditional probability of A, given B, is defined by:*

$$P(A|B) = \frac{P(A \cap B)}{P(B)}. \tag{1.2}$$

Often this equation is refound in a rule for calculating probabilities:

$$P(A \cap B) = P(B)P(A|B) = P(A)P(B|A). \tag{1.3}$$

This way to seemingly confuse a definition and a calculating rule may seem logically unsatisfactory, and it can only be justified through the fact that the definition (1.2) can be supported intuitively by Venn diagram considerations or by a range of examples of the type:

Example 1.9.1. Sample a person from a population where 50% are women, 30% are employed in a firm B and 20% are both women and employed in the firm B. Then, given the information that this person is employed in firm B, the probability that he/she is a woman will be 2/3. The example may be even more motivating if actual numbers are inserted for the percentages.

Two events A and B are called independent if $P(B|A) = P(B)$, or equivalently $P(A \cap B) = P(A)P(B)$. More generally:

Definition 1.9.2. *The events $A_1, A_2, \ldots, A_n$ are called independent if*

$$P(A_{k_1} \cap A_{k_2} \cap \ldots \cap A_{k_r}) = P(A_{k_1})P(A_{k_2}) \ldots P(A_{k_r}), \tag{1.4}$$

for all subsets of events.

It is clear from the axioms above and in the Appendix that the σ-algebra $\mathcal{F}$ in principle can be chosen in many different ways. In the real line case we can in principle let $\mathcal{F}$ consist of all sets, which turns out not to be practical in moderately complicated cases. A more useful option is then to let $\mathcal{F}$ consist of the Borel sets, the $\sigma-$algebra generated by the open subsets of the line. In practical applications this is not a big issue, and we will not focus at all on this problem in the present book. The reason is that the probability space Ω in many applications also is a topological space - often a subspace of some Euclidean space - , and then it is natural to choose $\mathcal{F}$ as the Borel σ-algebra, that is, the smallest σ-algebra containing all open sets. The reason is that a probability measure first can be defined on cartesian products of open intervals, then by a limiting operation on open sets. Or in the extremely simplest case: If Ω is finite or countable, we can take $\mathcal{F}$ to consist of all subsets.

From these rather innocent looking axioms a rather rich theory results. Several good books exist in probability; one can still recommend the classical ones by Feller [79, 80], or instance. Also, the theory of probability is the basis for statistical theory, and then also indirectly for all applications of this. The axiom system is rich enough to initiate an extensive asymptotic theory, which is of importance for several aspects of mathematical statistics. Since we in this book will mostly concentrate on statistical theory for a few experimental units, these aspects will be of no importance to us, however.

A final important concept is that of a random variable. Formally,

Definition 1.9.3. *A random variable $y = y(\cdot)$ is a function from the probability space Ω to the real line which is measurable, i.e., such that the inverse image $y^{-1}(B)$ of every Borel set B on the real line belongs to $\mathcal{F}$.*

The important aspect of this definition is that it enables us to calculate a probability distribution for every random variable and joint distributions for several random variables. Intuitively, we can think of a random variable as an observable quantity connected to some experiment which has a probability distribution, and this will be the important aspect to us.

Definition 1.9.4. *The probability distribution of a random variable y can always be specified through the distribution function F, which is a function of a real variable u given by*

$$F(u) = F_y(u) = P(y(\cdot) \leq u) = P(y^{-1}((-\infty, u])). \tag{1.5}$$

Definition 1.9.5. *a) If F is absolutely continuous with respect to Lebesgue measure du, it has a density $f(u) = F'(u)$, and we say that y is a continuous random variable, or that it has a continuous distribution.*

b) If F is constant except for a finite or countable number of jumps, we say that y has a discrete distribution, which then is specified by the frequency function $p(u)$, given by the successive sizes of these jumps.

In theory, any distribution function F which is non-decreasing fro all x and such that $F(x)$ belongs to the interval between 0 and 1 defines a probability measure. But in practice nearly all applications of probability are limited to either discrete or continuous distributions. The distribution of y is most often visualized using $f = f_y$ or $p = p_y$. One always has

$$\int f(u)du = 1 \quad \text{or} \quad \sum p(u) = 1. \tag{1.6}$$

Definition 1.9.6. *a) The expectation or expected value of the random variable y is given by*

$$\mu_y = E(y) = \int u f_y(u)du \quad \text{or} \quad \sum u p_y(u), \tag{1.7}$$

which graphically can be thought about as the center of gravity of the distribution.

b) The standard deviation $\sigma = \sigma_y$ is defined through the variance

$$\sigma_y^2 = Var(y) = \int (u - \mu_y)^2 f_y(u)du \quad \text{or} \quad \sum (u - \mu_y)^2 p_y(u). \tag{1.8}$$

Both standard deviation and variance are used as measures of the spread of the distribution. The advantage of the standard deviation is that it has the same unit as the random variable itself.

Example 1.9.2. Three important examples of continuous distributions are first the *normal distribution* with density

$$f_y(u) = \frac{1}{\sqrt{2\pi}\sigma} \exp(-\frac{(u-\mu)^2}{2\sigma^2}), \tag{1.9}$$

the *exponential distribution* with density

$$f_y(u) = \theta \cdot \exp(-\theta u), \quad u \geq 0, \tag{1.10}$$

and the uniform distribution with density

$$f_y(u) = \frac{1}{\beta - \alpha}, \quad \alpha \leq u \leq \beta. \tag{1.11}$$

The normal distribution (two parameters μ and σ) is widely used in models involving measurements, the exponential distribution (one parameter θ) in life time models for objects as different as bacteria and radioactive particles, while the uniform distribution is used in Monte Carlo simulations.

Example 1.9.3. Two important examples of discrete distributions are first the *binomial distribution*

$$p_y(u) = \binom{n}{u} \theta^u (1-\theta)^{n-u}, \quad u = 0, 1, 2, \ldots, n, \tag{1.12}$$

and then the *Poisson distribution*

$$p_y(u) = \frac{1}{u!} \theta^u e^{-\theta}, \quad u = 0, 1, 2, \ldots. \tag{1.13}$$

Both these distributions are used in biological modelling, for instance.

Note that all these classes of distributions depend upon one or several (here one or two) parameters. These parameters will be useful in the sequel. We will return to the applications in the next section and in Chapters 4 and 5.

A sequence of random variables $(y_1, y_2, \ldots)$ or a family of random variables $(y_t; t \geq 0)$ indexed by a continuous parameter t is called a stochastic process. Many statistical applications are based upon a finite collection $(y_1, y_2, \ldots, y_n)$ of random variables. Their joint distribution is then specified by the multivariate distribution function

$$F_{y_1, y_2, \ldots, y_n}(u_1, u_2, \ldots, u_n) = P(y_1 \leq u_1, y_2 \leq u_2, \ldots, y_n \leq u_n). \tag{1.14}$$

The corresponding multivariate density function, respectively frequency function can be defined from this in a relatively straightforward way.

If $y = (y_1, y_2, \ldots, y_n)$ is connected to an experiment, it is convenient to introduce a probability measure Q corresponding to the *multivariate distribution function* on the n-dimensional observation space/ sample space. Then

$$Q(A) = P(y^{-1}(A)) = P(\{\omega : y(\omega) \in A\}) = P(y \in A) \qquad (1.15)$$

for all measurable subsets A of $\mathbf{R}^n$. The specification of Q is equivalent to the specification of a multivariate distribution function F, which again is equivalent to the specification of a multivariate density or a multivariate frequency function; the latter will later in both cases be denoted by q.

A final important concept is that of the *conditional distribution* of the random variable y given another random variable x. From the joint and marginal distributions, the conditional density function or frequency function are defined by the following formulae, direct generalizations of (1.2):

$$f_{y|x}(v|u) = \frac{f_{x,y}(u,v)}{f_x(u)}, \quad p_{y|x}(v|u) = \frac{p_{x,y}(u,v)}{p_x(u)}. \qquad (1.16)$$

From this the conditional expectations and variances are defined in a straightforward way.

1.10 Probability Models for Experiments

Suppose now that one particular experiment $\mathcal{E} = \mathcal{E}^a$ on some system has been chosen, and that this experiment implies that a particular parameter $\theta^a = \theta$ is focused upon. This may be a scalar or a vector. We will assume that this parameter has a value, a statement which has a meaning which may be connected to this particular experiment.

This is in fact the standard situation for many statistical experiments. The word 'parameter' is traditional in statistics, and can best be understood when we look at the standard parametrized classes of distributions (1.9)–(1.13). However, we want to stress that in practical applications, parameter θ should usually be chosen first by carefully investigating the concrete situation, and then in most cases the probability model should be chosen afterwards. A parameter is a special case of a conceptually defined variable or c-variable. In fact, perhaps the latter is a better term in our setting, especially since the word parameter has different connotations, say, to physicists it may denote fixed physical quantities like the speed of light or the fine structure constant. However, it is difficult to alter a firmly established practice from statistics, so in agreement with the introduction we keep the word 'parameter' for a conceptually defined variable.

Imagine now the concrete experiment. It will be result in a set of observations collected in the (usually) multidimensional variable y. This may depend upon the state of the system under consideration, and it may also depend upon upon the chosen measurement apparata. In addition, some random noise is unavoidable in most practical experiments. All this is collected in the *model* of the experiment, which specifies a probability distribution for y:

$$Q^\theta(\cdot) = P^\theta(y \in \cdot). \tag{1.17}$$

The fact that this probability measure depends upon the state of the system through the parameter θ is crucial for all statistical inference. A large number of books have been written on statistical inference, and a summary will be written in Chapter 2 below.

Following up the assumption that θ has a value for the experiment, we will assume in this context that the parameter is identifiable under the experiment:

Definition 1.10.1. *The parameter θ is identifiable under the experiment corresponding to the model Q^θ if*

$$\theta_1 \neq \theta_2 \text{ implies } Q^{\theta_1} \neq Q^{\theta_2} \text{ for any } \theta_1, \theta_2. \tag{1.18}$$

This weak requirement is a prerequisite for all statistical inference. In fact, if θ is non-identifiable in all conceivable experiments, this comes very close to saying that it does not have a value. Then there is no experiment which can give a value to the parameter, that is, estimate the parameter uniquely from data.

The choice of the statistical model is a very important part of the inference process. A model is nearly always approximate, but approximate models can work surprisingly well. This is the reason why it is often sensible to use standard classes of distributions of the kind mentioned in Section 1.9 or their multivariate analogues to formulate models of experiments. These standard models often provide very accurate approximations.

Example 1.10.1. Let us discuss a simple example. In the population of potential voters in some democratic country there is some unknown number of voters who have decided to vote for some specific political party A. We make an opinion poll by selecting n voters randomly (in principle; in practice this is not easy) from the population. Let y be the random observation: The number of selected voters who say that they have decided for A.

We must make a number of assumptions of the type that the voters who say that they have decided to vote for A actually have made this decision. We must also assume that the total number N in the population and the unknown number N_A of potential A-voters are large compared to n. The unknown parameter or state variable is $\theta = N_A/N$. More complicated models can also be proposed, but they will need many paramaters, and thus be difficult to identify. Under the assumptions above we have the following situation:

1. There are n simple trials (selected voters) who each can take one of two values, A or not A. These trials are called simple since they only take two values.

2. There is a constant probability θ for A in each such trial.

3. The trials are independent.

It is an easy exercise to show (see Appendix A.1.2) that these conditions leads to the fact that y gets a binomial distribution with parameters n and θ. Thus the measure Q^θ given by the binomial frequency function (1.12) is our statistical model in this example.

The paradigm sketched above has been developed extensively by theoretical statisticians, and has been repeated by many textbooks: The concept of experiment is seemingly identical to that of a class of probability measures Q^θ indexed by some parameter θ. That this cannot be the case, was recently argued for by McCullagh [156], who showed by examples that some such 'models' were really absurd. A simple example is when one formula is chosen for the case when the number of observations is even, another formula when the number of observations is odd. In the vast majority of cases, such a changing model formula will lead to a very strange model.

McCullagh also developed a criterion, using category theory, to ensure that a 'model' is not absurd. In Chapter 3 we will come briefly back to this theme for the simpler case of submodels when a symmetry group is adjoined to the basic model.

The model concept, as it has been developed up to not, is a probability model of the data, given the parameters. A second application of probability theory to statistics and to natural sciences, in fact an application which later will be very important for us, is the following: In many cases it is possible to assume a probability distribution over the parameter or state variable itself. As a first item, we can often assume such a distribution prior to any experiment. Such a prior can be induced from previous knowledge, and can then be rather sharp, or it can be connected to ignorance about the parameter, and is then very diffuse.

In concrete applications the noninformative prior can be made up in several ways, but in this book — see Chapter 3 — I will introduce such a prior from the invariant measure of a symmetry group attached to the problem at hand.

The existence of a prior for the parameter for an experiment implies that we can use a kind of inference which conceptually is very simple, but which in concrete cases can be difficult to implement due to heavy computations. This is the so-called Bayesian inference, which is developed from Bayes' formula, in principle an extension of the conditional formula (1.2). The details of this will be explained in Section 2.6. Bayesian inference plays an increasing rôle in modern statistics, and several researchers concerned with the interpretation of quantum mechanics now also advocate viewpoints inspired by Bayesianism.

In classical physics probability distributions over c-variables are very important, in fact they lie behind the whole field of statistical mechanics. Later in this book we will also use such distributions in our development of quantum mechanics.

1.11   Elements of Group Theory

The concept of *group* is very important in all areas where symmetry considerations occur. In the setting of this book, every group will be a group of transformations, either of the parameter space or the space of observations. We will be interested in properties which transform in a reasonable way under such transformations: If an experiment leads to a conclusion about a parameter λ, then one might also be able to say something about the transformed parameter λg.

In mathematical terms, a group is defined as a collection G of elements g and an operation from $G \times G$ to G which satisfies:

Axiom 1.11.1. 1. Closure: $g_1 g_2$ exists for all g_1, g_2.

2. The associative law: $(g_1 g_2)g_3 = g_1(g_2 g_3)$ for all $g_1, g_2, g_3 \in G$.

3. A unit, or identity element: There exists a unique element e in G such that $eg = ge = g$ for all $g \in G$.

4. Inverse: For every element $g \in G$ there is a unique element g^{-1} for which $gg^{-1} = g^{-1}g = e$.

A group is called commutative (or abelian) if $g_1 g_2 = g_2 g_1$ for every $g_1, g_2 \in G$.

For our purposes these abstract axioms may have a somewhat limited interest except for the fact that they have certain mathematical consequences and describe mathematical objects which arise naturally in very many contexts. Most of our groups will be *transformation groups* constructed in the following way: Take as point of departure a set Θ, for instance a parameter space, then define a transformation on this space g by the property that for every point $\theta \in \Theta$ there is a unique new element $\theta g \in \Theta$. This transformation shall have the following properties:

$$\theta e = \theta$$

$$(\theta g_1)g_2 = \theta(g_1 g_2).$$

In many practical applications, such a g will correspond in some way or other to a symmetry of the set Θ.

Nevertheless, at some instances it will be useful to take as a basic object an abstract group G, that is, a set of elements together with its multiplication law. A transformation group on some set Θ which has elements and composition law inherited from G will then be called a set of *group actions*, in agreement with the use of words in pure mathematics. Throughout this book we will use the convention of placing the group element to the right of the object on which it acts. This convention will turn out to be useful later when defining invariant measures.

Any set of transformations which is closed under composition (i.e., such that $\theta(g_1 g_2)$ exists for all $\theta, g_1, g_2,$) will satisfy the associative law. This can be seen by using the second property in the definition of transformation above. A unit of the transformation set is given by the identity transformation, and an inverse is given by the opposite transformation, assumed to exist: If $\theta g = \theta'$, then $\theta' g^{-1} = \theta$. Hence such a class of transformations will always constitute a group, and satisfy the axioms above.

Here are some examples of transformation groups:

Example 1.11.1. Assume a location parameter μ and a scale parameter σ, so that $\theta = (\mu, \sigma)$.

a) A location group is given by a collection G of transformation group elements g_a defined by $(\mu, \sigma) g_a = (\mu + a, \sigma)$, or equivalently $\mu g = \mu + a, \sigma g = \sigma$.

b) A scale group is given by the transformations g_b such that $(\mu, \sigma) g_b = (b\mu, b\sigma)$, where $b > 0$.

c) A location and scale group is given by $(\mu, \sigma) g_{a,b} = (a + b\mu, b\sigma)$ with $b > 0$.

Example 1.11.2. If Θ is the space of n-dimensional real-valued vectors, then the affine group is defined by $\theta g = A\theta + b$, where g corresponds to some (A, b), where b is an n-vector and A is a non-singular $n \times n$ matrix. Similarly, the linear group is given by $\theta g = A\theta$, and the rotation group or orthogonal group by $\theta g = C\theta$, where the C's are orthogonal matrices ($C'C = I$, the identity, and $\det(C) = +1$).

In a similar way, if Θ is the space of complex-valued n-dimensional vectors, then the unitary group is given by $\theta g = U\theta$ with G corresponding to the space of unitary matrices ($U^\dagger U = I$, where $U^\dagger$ is defined by transposing and complex conjugation).

All these transformation groups generalize to the infinite-dimensional case, where A, C and U are suitably defined operators.

Example 1.11.3. If Θ is a finite set of, say d, parameter values, we can let G consist of all permutations of the elements θ. This is called the permutation group of order d. In general, a finite group is one with a finite number of elements. Finite groups have been extensively studied and classified by pure mathematicians. One can show that every finite group is a subgroup of a suitable permutation group.

The concept of *subgroup* will be important also for us. As the name suggests:

Definition 1.11.1. *A subgroup H of a group G is a subset of elements of G which itself constitutes a group. It is necessary that H is closed under composition, contains the unit e and contains the inverse of every element of H. The associative law is automatically inherited from G.*

So far we have looked upon transformation groups as transforming a parameter space. Later, see Section 3.2, it will also be useful to have a transformation group on the space of observations, and we will show that these two groups of transformations can be closely related. But first we will have a look on the transformation group defined on the inaccessible c-variable space, which also later will be important for us.

Example 1.11.4. Look at Example 1.3.3. Even if only one parameter from the c-variable $\phi = (\mu, \sigma)$ can be estimated, both the location group, the scale group and the location and scale group (see Example 1.11.1 above) are meaningful when defined in a straightforward way on this c-varable.

Example 1.11.5. Consider Example 1.3.4, where $\phi = (\theta^1, \theta^2)$ are the expected recovery times under two different treatments for one and the same patient. Let the time scale group be defined by $(\theta^1, \theta^2)g = (b\theta^1, b\theta^2)$. Then this group is defined and meaningful even though the vector parameter ϕ will not necessarily take a value.

Later, in our approach to quantum mechanics in Chapter 5, we will let the group actions at the outset be defined on the c-variable space Φ, and it is then important to observe from examples that these group actions can have a meaning even though the elements ϕ themselves do not take a value. Roughly, the reason is that ϕ can be defined through a description in some language, and relative to this description one can often easily imagine meaningful group actions.

One final important concept related to transformation groups is that of an *orbit*.

Definition 1.11.2. *Let again the group act on a parameter space Θ, and fix $\theta_0 \in \Theta$. The orbit containing θ_0 is then defined as the set of all parameters which can be written on the form $\theta_0 g$. In words, we follow the transformations starting from a fixed point. Any point on the orbit can be taken as starting point.*

When going from a group to a subgroup, each orbit will split up into several suborbits. Thus a large transformation group on a fixed space may have a small number of orbits, while, if going to a subgroup, the number of orbits increases.

Definition 1.11.3 *If there is only one orbit, we say that the transformation group is transitive. Then every element can be reached from every other element of the space by a group transformation.*

Further properties of group transformations will be discussed in Chapter 3, and some more technical aspects are given in Appendix A.2.

Chapter 2

STATISTICAL THEORY AND PRACTICE

2.1 Historical Development of Statistics as a Science

At all times people have used observations to make inferences about the state of nature. Of the pioneers pointing the way towards modern statistics and data analysis we will mention Jakob Bernoulli (1654-1705), Abraham de Moivre (1667-1754), Thomas Bayes (1702-1761), Leonhard Euler (1707-1783), Pierre S. Laplace (1749-1827), Carl F. Gauss (1777-1855), Lembert A.J. Quetelet (1796-1874) and Francis Galton (1822-1911); more details about these and other pioneers can be found in [141].

In 1895, Karl Pearson (1857-1936) recognized the need for more theoretical distributions than the normal one, and obtained his system of density functions as the solution of a certain differential equation. At the same time several new achievements were made: Collection and reduction of much empirical data; definition of the joint normal distribution and the corresponding coefficient of correlation; chisquare tests for goodness of fit; analysis of contingency tables and so on. About at the same time, mathematicians like Jules Henri Poincaré (1854-1912) were making contributions in the direction of adding rigor to the mathematical treatment of data.

All Pearsonian methods were only applicable to large samples. William S. Gosset (1876-1937) developed the socalled Student's t-test, which also is valid for small samples, a test which is very much used in practice today. Gosset cooperated closely with the famous statistician and geneticist Ronald A. Fisher (1890-1962), of many considered to be the father of modern statistics.

Fisher wrote a series of important articles on the general theory of estimation and inference in the 1920's; see [82]. Later, he developed the theory of analysis of variance and of the design of experiments [83, 84]. Fisher's contributions to statistics were diverse and fundamental. He developed the theory of maximum likelihood estimation, and introduced the important concept of sufficiency and ancillarity. Fisher was primarily concerned with the small sample of observations available from scientific experiments, and was careful to draw a sharp distinction between sample statistics (estimates) and population values (parameters to be estimated).

A mathematically clear and concise theory of statistical testing and confidence intervals was developed in the 1930's by Jerzy Neyman (1894-1981) and Egon S. Pearson (1895-1980). Early significance tests had been about the difference between binomial variables and between means, and they had been extended to the multivariate case by K. Pearson with the chisquare test and by Fisher with the analysis of variance test. Neyman and E. S. Pearson saw that such tests, to be effective, must consider the alternative hypothesis against which the null hypohesis was tested. They set out the two kinds of error in such testing, and thus arrived at their fundamental lemma for finding the optimal test in simple situations; they introduced the likelihood ratio test and the notion of power.

Modern statistics is characterized by several parallel developments: The field is being increasingly mathematical, and the gap between applied statistics and theoretical statistics has to some extent been widened. Yet statistical methods are being applied to an increasing number of areas, and many new branches of specializations and applications are being developed. The rôle of models in statistical and scientific work has become generally recognized, but it has also been pointed out that the model to some extent may be aestetical and arbitrary, even in situations where it in principle is based on experience and knowledge of the field of application. Bayesian methods (see Section 2.5 below) are being increasingly used, and multivariate methods are being more and more employed. Finally, the development of computers have made statistical methodology which were impossible to carry out decades ago, now quite feasible and common to use (bootstrap calculation, Markov Chain Monte Carlo calculations in Bayesian statistics, modern multivariate methods).

Still, there are many scientific areas where statistical methods are not used at all, and other areas where statistical methodology has to be supplemented by other ways of thinking. In psychology there is currently an active debate between the use of quantitative (statistical) and qualitative methods (see for instance [168] and references there). Though the use of statistics is recognizably useful in many applications, it is sometimes felt that more insight is being provided by case studies involving only a few individuals, but where each of these are thoroughly investigated. An interesting empirical study from medicine, where quantitative and qualitative methods are integrated is reported by Clarke [43]: Here, in a study of the effect of stroke, multiple regression on data from a large sample is used in the quantitative part, while, as a complementary part, focused interviews on 8 selected persons are used to shed light on the underlying reasons why and how factors operate to affect well-being following a stroke. Some traditional empirical researchers in psychology seem to use quantitative methods exclusively, while O'Neill [168] argues for an exclusive use of qualitative methods, at least in some cases. The present book, while mostly concentrating on other sciences than psychology, argues in general for a synthesis of methodologies; see also Chapter 9.

In a similar way as in psychology, verbal discussions through suitable concept definitions and with case studies to illustrate these concepts are very often found useful in sociological and political sciences, partly as an alternative to doing statistical investigations. The sociological reseach philosophy called positivism goes back to Auguste Comte (1798-1857) and has a quite rich content. In the last decades, however, it has been taken as synonymous with an attitude where methods from natural sciences, in particular statistical methods, are the only ones permissible, and has as such been severely critizised by many social scientists.

One obvious reason for this state of affair is that both psychology and social sciences by their nature are very complex compared to most natural sciences. A unit in psychology, most often a human being, can certainly often, at least from a certain point of view, be characterized by some state concept, either momentarily or more permanently. However, such a state is very frequently much more complex than what can be determined by some scalar or multivariate parameter, which standard statistical theory requires. In fact, complementary aspects are sometimes required in order to give a full description of phenomena. It is interesting that both psychologists and social scientists often use fixed schematic terms in characterizing humans. When a psychologist for instance characterize a client as being neurotic or schizophrenic, this definitively implies a model reduction relative to a more accurate state description of the client in most cases. I will argue later that model reductions may be useful, but a prerequisite is that they are consistent with relevant observable data.

2.2   The Starting Point of Statistical Theory

The statistical theory which follows in this chapter can be found in many different books, some of it in quite elementary books. On the intermediate level one can mention Freund [89] and Rice [177]. More advanced books are Berger [24], Bickel and Doksum [27] and Lehmann and Casella [146]. I have chosen in this chapter to go through from scratch all the necessary arguments from basic statistical theory, even those elementary aspects that are well known to statisticians. The purpose of this is twofold: First, and most important, I want to give the physicists and mathematicians among the readers the extra background needed to follow the arguments later in this book. But also I find it useful to go through the basic statistical theory again in the light of the discussion in Chapter 1. Comments to this end are mostly given in some of the examples, but also for instance in some of the discussion of the present Section. I find it important that this rethinking is carried out, not necessarily on a deep mathematical level, but at least to some degree at the level on which empirical scientists think. I think that even statisticians will be interested in the discussions given later in the chapter, but the earlier sections should also be looked at briefly.

As indicated in Section 1.9, the starting point for standard statistical inference is the statistical model $Q^\theta(\cdot) = P^\theta(y \in \cdot)$, the probability model for the observations as a function of the state of nature, the population parameter. Or more completely:

Definition 2.2.1. *The* standard statistical model *is defined by a sample space* $S \subset \mathbf{R}^n$, *the σ-algebra (A.1.1) $\mathcal{S}$ of Borel-subsets of S, and a family $\{Q^\theta(\cdot); \theta \in \Theta\}$ of probability measures on $(S, \mathcal{S})$, indexed by some parameter space Θ.*

The parameter θ can be one-dimensional or multidimensional, but most statistical methods require that the number n of units investigated shall be larger than the dimension p of θ. Later we will indicate how this requirement sometimes can be modified by using model reduction under symmetry.

An example of a statistical model was given in Example 1.9.1: The binomial distribution used to analyse an opinion poll. Typically, such a poll is more complicated: Even a single question requires more than two alternatives in the answer.

Example 2.2.1. Assume that n potential voters are picked out randomly, and that they are given to choose between $k-1$ political parties and the k'th alternative: 'don't know'. Let ζ_j be the probability that a single voter shall vote for party A_j; $j = 1, 2, \ldots, k-1$, and define $\zeta_k = 1 - \sum_{j=1}^{k-1} \zeta_j$, the probability of answering A_k, i.e., 'don't know'. Assume that the voters are independent, and let y_j be the number of voters in the sample who says that their vote will be A_j $(j = 1, 2, \ldots, k)$. By a straightforward extension of the argument leading to the binomial distribution, the joint distribution of $y_1, \ldots, y_k$ will be multinomially dependent upon the parameter $\theta = (\zeta_1, \ldots, \zeta_k)$. This means that the joint frequency function is:

$$P^\theta(y_1 = s_1, y_2 = s_2, \ldots, y_k = s_k) = \frac{n!}{s_1! s_2! \ldots s_k!} \zeta_1^{s_1} \zeta_2^{s_2}, \ldots, \zeta_k^{s_k}, \qquad (2.1)$$

where it is assumed that $s_1 + s_2 + \ldots + s_k = n$.

A natural estimate of ζ_j is y_j/n.

Also, of course, in most polls more than one question is being posed.

Example 2.2.2. Let the situation be as in Example 2.2.1, but with r questions in the poll. In question number i $(i = 1, 2, \ldots, r)$, let there be k_i alternative answers. Let ζ_{ij} be the probability of answering A_{ij} $(j = 1, 2, \ldots, k_i)$ on question i, and assume that y_{ij} of the voters actually answer A_{ij} on this question. (Note that then $\sum_{j=1}^{k_i} y_{ij} = n$ for each i.)

Then as before, under independence assumptions, each vector $(y_{i1}, y_{i2}, \ldots, y_{ik_i})$ is multinomial with parameter $\eta_i = (\zeta_{i1}, \zeta_{i2}, \ldots, \zeta_{ik_i})$, and the natural estimate of ζ_{ij} is y_{ij}/n. I will not here try to model explicitly the joint distribution of the y_{ij} for different i, but just mention that it can be found by generalizing the argument leading to the multinomial distribution for a single question.

But note that this is a situation with many parameters:

$$\theta = (\eta_1, \ldots, \eta_r) = (\zeta_{11}, \ldots, \zeta_{1k_1}, \ldots, \zeta_{r1}, \ldots, \zeta_{rk_r}). \qquad (2.2)$$

In fact the number of parameters here may even exceed n if you have a big questionnaire. The reason why it is possible to live with that in such a situation, is that one usually only analyses one question at a time, or just a few questions jointly in such questionnaires.

It is also important to go beyond the parameter concept to that of an inaccessible c-variable and to related concepts. The following example gives a small beginning to that.

Example 2.2.3. Assume that in May 2007 there is carried out an opinion poll among American voters concerning the question: 'Should the national car industry be encouraged by the state to produce less polluting cars?'; with the answer alternatives 'yes', 'no' or 'don't know'. Then among the respondents, some supply extra comments like 'I will decide right after gaining more experience on my own low-polluting car'; 'I have just seen Al Gore's documentary, and this made me change my opinion'; 'I will wait for further agreement among scientists on global warming'; 'It may seem to be an idea, but I don't know enough about the economical consequences'.

At the outset, this was designed as a simple poll with a two-dimensional parameter $\theta = (\zeta_1, \zeta_2, \zeta_3)$ (since $\zeta_3 = 1 - \zeta_1 - \zeta_2$). The extra information is vague, unsystematic and may not have been intended for by the researcher behind the poll. In fact, the interpretation of these comments may depend on whether the voters were asked to give comments or if the comments were an extra initiative shown by a few of the voters. In any case the comments may give some information, but in the latter case this may be regarded as some unexpected extra information in addition to θ. If it should be attempted to be systematized at all, this should be in terms of extra parameters $\theta^2, \theta^3, \ldots$, and then $\phi = (\theta, \theta^2, \theta^3, \ldots)$, in a related way as an inaccessible c-variable (Section 1.3) does not have a value in the original experiment.

I will give more examples of this kind in the following Sections, some of which will be much closer to the situation in quantum mechanics. But before that, I need to develop some more statistical theory.

Definition 2.2.1. *An* experiment *is any planned endeavour in order to get information about an object by means of observations. The object can be a population of units or a physical, biological or sociological system. The information that is sought is in terms of a number of questions, and the theoretical answers of those questions are characterizing the* parameters *of the experiment. In a standard statistical experiment these parameters will be assumed to have values, but in general a parameter - or more generally a c-variable - may or may not have a value relative to the given experiment.*

The vector of all parameters connected to an experiment $\mathcal{E}$ will be called the full
parameter θ *of that experiment. Several complementary experiments about the same
object may be performed, and the* c-variable $\phi = (\theta^1, \theta^2, \ldots)$, *where* θ^a *is the full
parameter of experiment* $\mathcal{E}^a$, *will as a rule not have a value relative to one given
experiment.*

*A c-variable which does not have a value relative to any conceivable experiment,
will be called* inaccessible.

Several examples of inaccessible c-variables were given in Section 1.3. In Example 2.2.3, the complementary information is strictly speaking not from a new
experiment, but it is at least information given outside the original, planned experiment. One feature of this example is that the parameter which is really of interest
to predict, is the outcome at the point of time when Congress later is to discuss
the issue, not the opinion status in May 2007. From the original experiment one
can only make such predictions under the simplified assumption that the division
of opinion will be constant during this period. The complementary comments may
serve to modify this simple assumption, or to design a new poll, where comments
of this kind are requested in a systematic way.

In the next example, there may be a similar problem. The example is first of all
given here, however, to illustrate a very common setting for a standard statistical
experiment, namely that of an independent, identically distributed series of observations, and the question of when these assumptions can be transferred to another
setting.

Example 2.2.4. From some hospital in Norway, a human population with a
certain disease is identified, and a random sample of n such patients are picked out.
Each of these is given a specific treatment which is under investigation. Assume for
simplicity that all patients recover, and the recovery times are given by $y_1, y_2, \ldots, y_n$.
Since the patients are independent, these can be assumed to be independent random
variables. The parameters of interest are the mean recovery time μ, and as a measure
of the variation of the recovery time the standard deviation σ.

As a model for the observations we then propose: $y_1, y_2, \ldots, y_n$ are independently distributed, each having a normal distribution (1.9) with expectation μ and
standard deviation σ.

This is a simplification in most cases, but it will often be a useful simplification.
From this model we will develop simple statistical procedures in the next two ses-
sions. We will estimate the parameters, say something about the uncertainty of the
estimates, formulate confidence intervals and test hypotheses.

Assume now, however, that a group of patients with the same disease is found
in England. The problem is then: Can the results from the investigation above be
used for the corresponding population? Such decisions have to be done every day
in medicine. In mathematical terms the question is: An investigation is carried
out for a certain population with parameters $\theta = (\mu, \sigma)$. In a related population

the parameters are $\theta' = (\mu', \sigma')$. Can one assume that $\theta = \theta'$? In many cases it seems reasonable to assume this as an approximation if the difference between the populations is unrelated to the disease in question and to variables which can be related to the disease, but it must be decided by arguments in each specific case.

2.3 Estimation Theory

In this and the following sections we will assume a standard statistical experiment with a parameter θ which has a fixed value. There is a vector of observations y, and a model for these observations given by $Q^\theta(\cdot) = P^\theta(y \in \cdot)$. The (multivariate) probability density or frequency function corresponding to this model is given by $q^\theta(u)$.

The first step of a statistical analysis may be to find an estimate $\hat{\theta}$ of the parameter θ or an estimate $\hat{\mu}$ of some component of θ. Such estimates were already provided in Example 2.2.1 above. Correspondingly, in Example 1.9.1, a poll with one question and two answer alternatives, the estimator will be $\hat{\theta} = y/n$. An estimate will in general be a function of the observations which estimate a parameter of interest. If the observations are regarded as random variables, we talk about the corresponding *estimator* $\hat{\theta}$ or $\hat{\mu}$.

Definition 2.3.1. *An estimator $\hat{\theta}$ is called* unbiased *if* $E(\hat{\theta}) = \theta$.

Of course it is a nice property for an estimator to be unbiased, but often it is enough to require approximate unbiasedness. This may be made precise in many ways, but often it is implemented as asymptotic unbiasedness, meaning that if the experiment is thought to be repeated many independent times, the hypothetical mean of the corresponding estimates will be about equal to the unknown parameter θ. In general, the *bias* of an estimator is defined as $E(\hat{\theta}) - \theta$.

Another property of interest is the spread of these hypothetical repeated measurement, which is found from the model of a single experiment by $Var(\hat{\theta})$. This variance is useful, but unfortunately, in most cases it depends upon the unknown parameter.

A measure which takes into account both the bias and the variance, is given by the mean square error:

$$ME(\hat{\theta}) = (E(\hat{\theta}) - \theta)^2 + Var(\hat{\theta}) = E(\hat{\theta} - \theta)^2, \tag{2.3}$$

where the last identity is shown by a simple expansion.

In the examples mentioned above, it was obvious what the estimator should be, but it is not always so in practical applications. There has been developed several systematic methods for finding good proposals for estimators. In principle these should always be evaluated from the model by using the criteria just mentioned, but this is not always done in practice.

(1). The method of moments

Suppose that the observations $y_1, y_2, \ldots, y_n$ are independent and identically distributed with a p-dimensional parameter $\theta = (\mu_1, \ldots, \mu_p)$. Let $\tau_j = \tau_j(\theta) = E^\theta(y_1^j)$, $j = 1, \ldots, p$. Then these are estimated by

$$\hat{\tau}_j = \frac{1}{n} \sum_{i=1}^n y_i^j. \tag{2.4}$$

The τ_j's can be found from the model as functions of θ, and in the case where a unique solution can be found, a natural proposals for $\hat{\theta} = (\hat{\mu}_1, \ldots, \hat{\mu}_p)$ is given by solving the equations

$$\hat{\tau}_j = \tau_j(\hat{\theta}), \ j = 1, 2, \ldots, p. \tag{2.5}$$

Example 2.3.1. As in Example 2.2.4 let the common distribution be normal (μ, σ), i.e., with density

$$f_{y_1}(u) = \frac{1}{\sqrt{2\pi}\sigma} \exp\left(-\frac{(u-\mu)^2}{2\sigma^2}\right).$$

This distribution has 2 parameters, and a simple integration gives

$$E(y_1) = \mu, \quad E(y_1^2) = \mu^2 + \sigma^2.$$

Thus the method of moments equations are:

$$\hat{\mu} = \bar{y} = \frac{1}{n} \sum_{i=1}^n y_i, \quad \hat{\mu}^2 + \hat{\sigma}^2 = \frac{1}{n} \sum_{i=1}^n y_i^2, \tag{2.6}$$

giving:

Definition 2.3.1.

$$\hat{\mu} = \bar{y}, \quad \hat{\sigma}^2 = \frac{1}{n} \sum_{i=1}^n (y_i - \bar{y})^2.$$

The estimator $\bar{y}$ for μ is much used and has several optimality properties. But usually, $\hat{\sigma}^2$ is replaced by the unbiased variance estimator

Definition 2.3.2.

$$s^2 = \frac{1}{n-1} \sum_{i=1}^n (y_i - \bar{y})^2. \tag{2.7}$$

Example 2.3.2. In applications of the uniform distribution (1.11) the parameters α and β are most often determined from the setting. However, in this example we assume a situation where they are to be estimated from independent data $y_1, \ldots, y_n$. A straightforward calculation gives

$$\tau_1 = E(y_1) = \frac{1}{2}(\alpha + \beta), \quad \tau_2 = E(y_1^2) = \frac{1}{3}(\alpha^2 + \alpha\beta + \beta^2).$$

Solution of this together with (2.4) gives a second degree equation for $\hat{\alpha}$ and $\hat{\beta}$.

(2). The least squares method

The method of least squares will be very important when we come to regression and analysis of variance. At present we just illustrate it by a simple example:

Example 2.3.3. Let the situation be as in Example 2.3.1, and let us implement the requirement that the estimator of μ shall be as close as possible to the center of the observed y_i's by saying that it shall minimize the least squares criterion

$$LS(\mu) = \sum(y_i - \mu)^2. \qquad (2.8)$$

A simple exercise shows that this gives $\hat{\mu} = \bar{y}$ again.

(3). Maximum likelihood

This method, which was proposed by Fisher, can be shown to have very good properties under reasonable assumptions for large samples, and has dominated much of statistical theory and practice for decades. It is in fact conceptually very simple, and gives good results in a number of applications, at least when the number of parameters is not too large.

Look at the model Q^θ for the data y and the corresponding density or frequency function $q^\theta(u)$. Replace here u by the data y, and consider this as a function of θ. This gives the *likelihood*:

$$L(\theta) = q^\theta(y). \qquad (2.9)$$

From a certain point of view, this function indicates how well a given parameter explains the data obtained: A large $L(\theta)$ means a high value of the probability density at the point where the actual data were found, which again can be taken as an indication that the density for this particular value of the parameter gives a good explanation of the data. Hence the maximum likelihood method consists simply of maximizing the function L to find a good estimation proposal $\hat{\theta}$. (Since L also depends upon y, this $\hat{\theta}$ will depend upon the observations y.)

In practice, it is often easier to maximize the *log likelihood*

$$l(\theta) = \ln L(\theta), \qquad (2.10)$$

a quantity which later will turn out also to have some useful theoretical significance.

Example 2.3.4. Assume again an independent, identically normally distributed series of observations $y_1, \ldots, y_n$, so that

$$L(\mu, \sigma) = \prod_{i=1}^{n} \frac{1}{\sigma\sqrt{2\pi}} \exp(-\frac{1}{2}(\frac{y_i - \mu}{\sigma})^2). \qquad (2.11)$$

From this

$$l(\mu, \sigma) = -n\ln\sigma - \frac{n}{2}\ln 2\pi - \frac{1}{2\sigma^2} \sum_{i=1}^{n}(y_i - \mu)^2. \tag{2.12}$$

A straightforward optimalization from this shows that

$$\hat{\mu} = \bar{y}, \quad \hat{\sigma}^2 = \frac{1}{n}\sum_{i=1}^{n}(y_i - \bar{y})^2, \tag{2.13}$$

the same as the method of moments gives in this case.

Example 2.3.5. In Example 1.9.1 we have one parameter θ and

$$L(\theta) = \binom{n}{y} \theta^y (1 - \theta)^{n-y}, \tag{2.14}$$

hence

$$l(\theta) = \text{const.} + y\ln(\theta) + (n - y)\ln(1 - \theta). \tag{2.15}$$

Optimalization again gives $\hat{\theta} = y/n$, the natural estimator. The argument can be generalized to the multinomial distribution.

Example 2.3.6. Consider again as in Example 2.3.2 the estimation of α and β in the uniform distribution (1.11). The likelihood here is $(\beta - \alpha)^{-n}$ when all y_i's are between α and β, otherwise 0. Maximizing the likelihood directly here gives

$$\hat{\alpha} = \min(y_i), \quad \hat{\beta} = \max(y_i),$$

that is, something completely different from the moment estimators in Example 2.3.2. The maximum likelihood stimators will obviously be biased, giving a slightly too small range $\hat{\beta} - \hat{\alpha}$. If they are multiplied by constants to make them unbiased, the resulting estimators will be quite good.

Under regularity conditions there is a nice asymptotic theory of maximum likelihood estimation relating it to several optimality properties; see [146], but very few general properties are known for the method for small samples. Thus for such samples the estimator must be investigated separately in each single case.

2.4   Confidence Intervals, Testing and Measures of Significance

Once an estimator is found, let us assume that it is unbiased or nearly unbiased, its uncertainty is given by its variance $Var(\hat{\theta})$. As already remarked, this variance is usually dependent upon the unknown parameter, but in many cases the variance can also be estimated from the available data, and then we have a concrete uncertainty measure given by $\widehat{Var}(\hat{\theta})$ or its square root. The estimated standard deviation $\sqrt{\widehat{Var}(\hat{\theta})}$ is often called the *standard error* of $\hat{\theta}$.

Example 2.4.1. Look again at the case with n independent normally distributed observations. A transformation exercise, see Appendix A.1.3, then shows that $\bar{y}$ is normally distributed with parameters μ and σ^2/n, and that $\bar{y}$ is independent of s^2. In particular then, the first of these informations gives

$$Var(\hat{\mu}) = Var(\bar{y}) = \frac{\sigma^2}{n}. \tag{2.16}$$

Furthermore, we know that in this case σ^2 is unbiasedly estimated by s^2, so a measure of uncertainty connected to $\hat{\mu}$ is given by s^2/n, or its square root $s/\sqrt{n}$. This result can be shown also to be valid for other distributions than the normal one (if one in general defines μ as the expectation of the distribution), and the result has great practical implications.

Beginning with J. Neyman and E. G. Pearson, statisticians have made results of this kind much more precise. Assume first a one-dimensional parameter θ with some estimator $\hat{\theta}$. Assume in addition that it is possible to find a *pivot*, that is, a function

$$C(\theta, \hat{\theta}) \tag{2.17}$$

whose distribution is independent of θ. Then, for a given probability γ, say 0.95, we can find constants c_1 and c_2 such that

$$P(c_1 \le C(\theta, \hat{\theta}) \le c_2) = \gamma.$$

If for each $\hat{\theta}$, $C(\theta, \hat{\theta})$ is continuous and monotonic in θ, this can be inverted to give

$$P(d_1(\hat{\theta}) \le \theta \le d_2(\hat{\theta})) = \gamma. \tag{2.18}$$

This is then a very practical and important result: There is an interval $[d_1(\hat{\theta}), d_2(\hat{\theta})]$, depending only upon the observations, so that, before the experiment is done, that is, essentially as a function of the estimation method used, there is a fixed probability γ that this interval shall contain the unknown parameter θ. This interval is called a confidence interval. If γ is 0.95, say, we talk about a 95% confidence interval.

Example 2.4.2. Consider again the normal measurement series $y_1, \ldots, y_n$, and assume now that the variance σ^2 is known, or can be assumed known from some given information. (This will then be a part of the context for the experiment.) Recall from Example 2.4.1 that $\bar{y}$ for such a measurement series also is normal. The normal distribution turns out to transform the way one expects under translations and scale changes, so the pivot

$$\frac{\bar{y} - \mu}{\sigma}\sqrt{n} \tag{2.19}$$

has a *standard normal* distribution, a normal distribution with $\mu = 0$ and $\sigma = 1$. The corresponding distribution function has been evaluated numerically. It is a well-known fact that $P(-1.96 \le u \le 1.96) = 0.95$ for a standard normal u, so

$$P(-1.96 \le \frac{\bar{y} - \mu}{\sigma}\sqrt{n} \le 1.96) = 0.95$$

giving

$$P(\bar{y} - 1.96\frac{\sigma}{\sqrt{n}} \leq \mu \leq \bar{y} + 1.96\frac{\sigma}{\sqrt{n}}) = 0.95.$$

From this, we conclude that a 95% confidence interval for μ is given by

$$[\bar{y} - 1.96\frac{\sigma}{\sqrt{n}}, \bar{y} + 1.96\frac{\sigma}{\sqrt{n}}]. \tag{2.20}$$

As a concrete numerical example, consider the following measurement series of ten observations: 850, 960, 880, 890, 890, 740, 940, 880, 810, 840. These are the first 10 in a longer series from Michelson's determinations of the velocity of light from 1879 as reported by Stigler [194] and Rice [177]. The measurements are in km/s, and the fixed number 299 000 has been subtracted. From these numbers we find $\bar{y} = 868$ and $s = \sqrt{s^2} = 59.8$ (cp. Definition 2.3.2). We will illustrate how to make use of the last number in the next Section. In the present case it turns out to be an underestimate of the variation; for the full series of 100 observations between June 5, 1879 and July 2, 1879 as reported in [177] we find $s = 78.6$, indicating that the real σ is about 80. Assuming now $\sigma = 80$ and returning to the 10 masurements above, we find that the 95% confidence interval for the expected measurement value is

$$[868 - 1.96\frac{80}{\sqrt{10}}, 868 + 1.96\frac{80}{\sqrt{10}}] = [818.4, 917.6].$$

From these values we conclude that Michelson's 10 measurements indicate a value of the speed of light betwen 299 818.4 km/s and 299 917.4 km/s. From modern measurements we know that the real value is 299 792.5 km/s, so Michelson was not far off. Nevertheless, we can conclude that Michelson's method was slightly biased?

(For completeness we should say that the full series of 100 measurements reported in [177] gives $\bar{y} = 852.4$ and a 95% confidence interval from 836.7 to 868.1 for the expected value. Compared to the modern value $\mu = 792.5$, it seems quite clear that Michelson was biased.)

The discussion in this example can then also be used to illustrate the second statistical method first discussed fully by J. Neyman and E. G. Pearson in the 1930's, namely that of *hypothesis testing*. We assume again first a series of measurements y with a model Q^θ depending upon a scalar parameter θ. Assume that there is a fixed value θ_0 of this parameter which is of particular interest, like the modern μ-value for the velocity of light in the example above. Typically, we are interested in using the data to decide between two hypotheses given as follows:

Definition 2.4.1. *The null hypothesis is given by* $H_0 : \theta = \theta_0$ *and the alternative hypothesis by* $H_1 : \theta \neq \theta_0$.

Usually, as in the Michelson example, the null hypothesis is the statement that we want to keep if the data do not clearly indicate something else, here that the measurement method used by Michelson was *not* biased.

To construct a test, we first find a *test variable*, a function of the data $t(y)$ which tends to be moderate under H_0 and large under H_1. There are systematic methods of finding good test variables, but we will only stick to examples here.

Definition 2.4.2. *The test procedure is: We reject H_0 and claim H_1 if the data give $t(y) > t_0$, where t_0 is a predetermined rejection constant.*

This rejection constant is determined from the basic requirement

$$P(t(y) > t_0|H_0) = \alpha, \tag{2.21}$$

where α is some predetermined small constant; traditional values are 0.05, 0.01 and 0.001. The constant α is called the level of the test.

A small level indicates that we are very strongly predetermined to stick to H_0. If the null hypothesis is rejected at the 0.001 level, this is a very strong indication that the alternative hypothesis indeed is true.

For completeness and for later reference we include a few more concepts related to testing.

Definition 2.4.3. *The power of a test is the probability of rejecting H_0 as a function of the parameter θ.*

In a very simple case, the test of a simple hypothesis against another simple hypothesis, the optimal solution is well known, as given by the Neyman-Pearson lemma ([144]).

Lemma 2.4.1. *Let the standard statistical model be given by a density $q^\theta(y)$, and assume that we want to test $H_0 : \theta = \theta_0$ against the alternative $\theta = \theta_1$. Then the test at a fixed level with the highest power is given by rejecting H_0 when $q^{\theta_1}(y)/q^{\theta_0}(y) > $ const., where the rejection constant is found from the level requirement.*

The usefulness of Neyman-Pearson's lemma is that it can be used to suggest good tests in other simple situations.

Example 2.4.3. Return to the 10 Michelson measurements from Example 2.4.2. We want to test the null hypothesis H_0: Michelson's method was not biased against H_1: The method was biased. In mathematical terms this means: $H_0 : \mu = \mu_0$ against $H_1 : \mu \neq \mu_0$, where μ is the expectation of Michelson's measurements and $\mu_0 = 792.5$.

As a test variable it is reasonable to choose $t(y) = |\bar{y} - \mu_0|$. As demanded, this is a quantity which tends to be small under H_0 and large under H_1. The distribution of $t(y)$ under H_0 is found from the fact that $\bar{y}$ is normal $(\mu_0, \sigma^2/n)$ under H_0 with

σ still assumed to have a known value of 80. If the test level is 0.05, the rejection constant is found from the fact that

$$P^{\mu_0}(|\bar{y} - \mu_0| > t_0) = 2P^{\mu_0}(\frac{\bar{y} - \mu_0}{\sigma}\sqrt{n} > \frac{t_0}{\sigma}\sqrt{n}) = 0.05$$

requires $t_0\sqrt{n}/\sigma = 1.96$ or $t_0 = 1.96\sigma/\sqrt{n} = 49.6$. Thus, H_0 is rejected if $|\bar{y} - \mu_0| > 49.6$. In the present case $|\bar{y} - \mu_0| = |868 - 792.5| = 75.5$; thus H_0 is rejected at the 5% level, and we conclude relative to this level that Michelson's measurements were biased.

It can be seen from the above algebra that the null hypothesis is rejected at the 5% level if and only if a 95% confidence interval does not cover the fixed constant μ_0. Thus the conclusion here could have been seen already from Example 2.4.2. Such a connection between confidence intervals and hypothesis testing is quite general. Nevertheless, a separate disussion of testing of hypotheses is very useful, and testing is in fact much used in practice. First, testing can be generalized to the one-sided case and to multivariate cases. Next, testing can be performed on many different levels using the same test variable.

In the present case, if the testing is done on the 1% level, the rejection constant will be 65.2, and at the 0.1% level it will be 83.5. Thus the null hypothesis is rejected at the 5% and 1% level, but not at the 0.1% level. If we want to be very safe when we conclude that the method is biased, we can not conclude that from these 10 data values. Using all the 100 data values, the null hypothesis is rejected on any reasonable level, however.

In applied work, testing is often done through the calculation of P-values. The P-value for a test is found as follows: Determine first the value of the test statistics actually found in the experiment, in the example above 75.5. Then find from the null distribution the probability that $t(y)$ shall be larger than that, which is called the P-value of the test and is reported as a probability or a percentage. In the example, P=0.26%. Since P is less than 1%, for instance, the null hypothesis is rejected at the 1% level as before.

Definition 2.4.4. *As an alternative to rejecting* H_0 *at a fixed level, report the P-value* $P = P(t(y) > t_{y_0}|H_0)$, *where* y_0 *is the fixed, observed data.*

In the way hypothesis testing is described, it is tacitly understood that the parameter really has a value. This may be close to the truth if there is a well-defined population underlying the collected observations. However, some care must be exercised. In the case where the null hypothesis is *not* rejected, it is far from correct in general to conclude that $\theta = \theta_0$. For one thing, further data may still give rejection of H_0. Furthermore, the model chosen is most often an approximation, and θ may not even make sense outside the model.

Nevertheless, putting one or several parameters equal to their null values after a non-rejecting test is often done in practice for instance in connection to multiple regression models; see later. In these cases, the new, simplified model is a deliberate choice, and is part of the context for the further investigations, not necessarily as an objective "truth" to be reported.

Here is a remark of relevance to the situation of elementary quantum mechanics, though: In this setting a parameter or state variable λ takes a number of discrete values, say, $\lambda_1, \ldots, \lambda_k$, ordered according to size. If one particular value first is used as a null hypothesis in a test, it is also possible to perform tests where each neighbouring value is used as a null hypothesis. Then, if the middle value is not rejected, and the neighbouring values are rejected, it is indeed possible to report the middle one as the truth as resulting from the experiment. Alternatively, if a 99.9% confidence interval only covers one of the parameter values, we can be reasonably sure that this is the correct one.

2.5　Simple Situations Where Statistics is Useful

Estimation, confidence interval estimation and testing of hypotheses is very often done in connection to one or several measurement series $y_1, .., y_n$ of independent identical observations. Often, these observations are assumed to be normally distributed, a simplification, but very often a useful simplification. I will come back to this under model reduction later. (See Example 3.1.5.) At this point it must only be repeated that every model is a simplification. In most practical situations it is probably not even correct to say that there exists a unique, true model, so by an extention of an earlier concept we might say that the variable 'model' does not have a value. Nevertheless, the concept of a model is a very useful one, and there are several good reasons for choosing a normal model when it is not invalidated by the data themselves. It is simple; it depends upon just the right two parameters μ and σ; it transforms in the right way under translations and scale changes; it may be argued for by the central limit theorem [80] in some cases; there exist a multitude of good statistical methods for analysing normal observations; and finally, these methods are in most cases reasonably robust against deviations from the normality assumption, at least in the case where this distribution is independently investigated by plots or in other ways.

In the previous section, we treated the case where the variance σ^2 of the observations was known. To treat the more general case, we need three auxiliary sets of distributions. Throughout, we will let $u, u_1, u_2, \ldots, u_n$ be independent, and each have a standard normal distribution, i.e., be normal with $\mu = 0$ and $\sigma = 1$.

Definition 2.5.1. The chisquare distribution. *Let*

$$z = u_1^2 + \ldots + u_\nu^2. \tag{2.22}$$

Then per definition z has a chisquare distribution with ν degrees of freedom.

Definition 2.5.2. The t-distribution. *Let*

$$t = \frac{u}{\sqrt{z}}\sqrt{\nu}, \tag{2.23}$$

where z has a chisquare distribution as in Definition 2.5.1 with ν degrees of freedom, and where u and z are independent. Then per definition t has a (Student's) t-distribution with ν degrees of freedom.

Definition 2.5.3. The F-distribution. *Let*

$$F = \frac{z_1 \nu_2}{z_2 \nu_1}, \tag{2.24}$$

where z_1 and z_2 are independent and chisquare distributed, z_1 with ν_1 degrees of freedom and z_2 with ν_2 degrees of freedom. Then per definition F has a (Fisher's) F-distribution with ν_1 and ν_2 degrees of freedom.

Tables related to these distributions are given in nearly all statistical textbooks, also in elementary books, and the densities of the distributions are found in more advanced books.

Example 2.5.1. In a single measurement series $y_1, \ldots, y_n$ where the measurements are independently normal (μ, σ), it follows from the result of Appendix A.1.3 and from Definition 2.5.2 that

$$\frac{\bar{y} - \mu}{s}\sqrt{n} \tag{2.25}$$

has a t-distribution with $n - 1$ degrees of freedom. From this, a 95% confidence interval for the parameter μ is given by

$$[\bar{y} - t_0 \frac{s}{\sqrt{n}}, \bar{y} + t_0 \frac{s}{\sqrt{n}}], \tag{2.26}$$

where t_0 is found from tables of the t-distribution with $n-1$ degrees of freedom such that $P(|t| > t_0) = 0.05$. A test on the 5% level of $H_0 : \mu = \mu_0$ against $H_1 : \mu \neq \mu_0$ is given by rejecting H_0 if

$$\frac{|\bar{y} - \mu_0|}{s}\sqrt{n} > t_0. \tag{2.27}$$

Example 2.5.2. Comparing two measurement series is something that is often done in simple experiments, both in biology and in medicine. Assume that two diets for slimming purpose are to be compared. From a certain target population, n_1 persons are picked out randomly and given diet 1 during a certain interval. Their weight losses during this period are $y_{11}, y_{12}, \ldots, y_{1n_1}$. Similarly, from the same population, n_2 persons are given diet 2, resulting in the weight losses $y_{21}, y_{22}, \ldots, y_{2n_2}$.

Assuming that the variances in the two populations are the same, the best pooled variance estimate can be shown to be

$$s^2 = \frac{(n_1 - 1)s_1^2 + (n_2 - 1)s_2^2}{n_1 + n_2 - 2}; \quad s_i^2 = \frac{1}{n_i - 1}\sum_{j=1}^{n_i}(y_{ij} - \bar{y}_{i\cdot})^2, \tag{2.28}$$

which is such that $(n_1 + n_2 - 2)s^2/\sigma^2$ has a chisquare distribution with $n_1 + n_2 - 2$ degrees of freedom. This then, similarly as in Example 2.5.1, leads to a t-distribution connected to the difference in means $\bar{y}_1. - \bar{y}_2.$, or more precisely to $t = (\bar{y}_1. - \mu_1 - \bar{y}_2. + \mu_2)/s\sqrt{n_1^{-1} + n_2^{-1}}$, where μ_1 and μ_2 are the two expected weight losses. From this we find in a straightforward way, similarly as in Example 2.5.1, a t-test for testing $H_0 : \mu_1 = \mu_2$ against $H_1 : \mu_1 \neq \mu_2$. The number of degrees of freedom is $n_1 + n_2 - 2$.

Even though this is done as a two-sided test, it is obvious that, if the null hypothesis is rejected, it is not only of interest to state this conclusion, but also which of the two diets that is claimed to give the highest expected weight loss.

Example 2.5.3. In Norway and also internationally there has indeed been a general debate concerning what kind of diet which is best for people that decide to slim, a low fat one based on total energy content or a low carbohydrate diet based on the so-called glycemic index. Investigations have been done to elucidate this question, and more investigations are being planned.

To simplify, consider the following two experimental schemes, where in both cases two samples are compared as in Example 2.5.2: 1) Each person is given a detailed diet list for each day during the experiment, where these diets are composed by experts in agreement with the two philosophies. 2) Each person is given a list of 'good' or 'bad' products according to the selected one of the two philosophies, and is required to do his shopping in agreement with this.

The point is then that the different experimental schemes really measures different properties, the last one also taking into account some of the psychological elements associated with slimming. If only one type of experiment is carried out, it may be difficult to evaluate the complementary property associated with the other experiment. In order to decide which experiment to do, one should consider which properties that are most important for practical advice.

We now turn to somewhat different classes of experiments, also often carried out in practice, namely those which are associated with the binomial and the multinomial distributions. We will only give a very brief survey.

Consider first the situation in Example 1.9.1, the binomial poll with one question and two possible answers. The probability θ of voting A is estimated by $\hat{\theta} = y/n$, where y is the number among the n persons in the poll which answer A. From the results of Appendix A.1.2 we find $E(\hat{\theta}) = \theta$ and $Var(\hat{\theta}) = \theta(1 - \theta)/n$. Using a normal approximation, which can be argued for by the central limit theorem ([80]), we find the following 95% confidence interval:

$$[\hat{\theta} - 1.96\sqrt{\frac{\hat{\theta}(1 - \hat{\theta})}{n}}, \hat{\theta} + 1.96\sqrt{\frac{\hat{\theta}(1 - \hat{\theta})}{n}}]. \qquad (2.29)$$

A 5% test of $H_0 : \theta = \theta_0$ against $H_1 : \theta \neq \theta_0$ (where θ_0 may be a previous election result), is given by rejecting H_0 if

$$\frac{|\hat{\theta} - \theta_0|}{\sqrt{\frac{\theta_0(1-\theta_0)}{n}}} > 1.96. \tag{2.30}$$

This latter inequality may also be written

$$z = \frac{(y - n\theta_0)^2}{n\theta_0} + \frac{((n-y) - n(1-\theta_0))^2}{n(1-\theta_0)} > 1.96^2. \tag{2.31}$$

This seemingly cumbersome way of writing a test variable in fact is a first indication of a great number of possible generalizations: In a multinomial situation or in a set of multinomial situations a given null hypothesis is tested by first calculating the expected number η_i under the null hypothesis in each cell, then the test variable

$$z = \sum_i \frac{(y_i - \eta_i)^2}{\eta_i}, \tag{2.32}$$

where y_i is the observation in cell i. Finally the null hypothesis is rejected if $z > z_0$ with z_0 calculated from the chisquare distribution with $s(q-1)$ degrees of freedom, s being the number of multinomials and q the number of cells per multinomial. Note that sq is the total number of cells. The test can also be modified by replacing the expected numbers η_i by some $\eta_i(\hat{\zeta})$, where $\hat{\zeta}$ is a (often multidimensional) parameter estimated under the null hypothesis (by maximum likelihood or minimum chisquare). The number of degrees of freedom must be replaced by $s(q-1) - k$, where k is the dimension of ζ. All these tests are approximate, more precisely asymptotically valid as q tends to infinity and $\hat{\zeta}$ tends in probability to ζ.

Example 2.5.4. Assume that a researcher is interested in if there is a difference in attitudes towards a certain problem between Norway and England. A question with r response alternatives $A_1, \ldots, A_r$ is formulated to investigate this, and the question is posed to n_1 randomly selected persons in Norway and to n_2 randomly selected persons in England. The number of persons answering A_i is y_{i1} in Norway and y_{i2} in England. The null hypothesis is that there is a common probability θ_i to answer A_i in England and Norway, and the estimate of this under the null hypothesis is $\hat{\theta}_i = (y_{i1} + y_{i2})/(n_1 + n_2)$. Thus the expected number in cell i in Norway is estimated by $\eta_i(\hat{\theta}) = n_1\hat{\theta}_i$, which is to be compared to y_{i1}. Similarly, in England $n_2\hat{\theta}_i$ is to be compared to y_{i2}. There are $r - 1$ independent parameters θ_i to estimate, and the number of degrees of freedom for the chisquare test is $2(r-1) - (r-1) = r - 1$.

The situation here can be generalized to several questions, and also to comparing more than two countries.

All examples of statistical procedures in this section have been for experiments where many units are investigated, and this is the typical situation in applied statistics. When we later come to quantum physics, we will be concerned with one or only a few units. In fact, this may be one reason why the conceptual similarity between statistics and quantum theory has not been focused on before. I will show nevertheless that a connection exists, but in order to see that, we have to carefully look at the different facets of the concept of parameter or state variable. In fact, as we saw in the Michelson example, repeated measurements are used also in physics.

2.6 Bayes' Formula and Bayesian Inference

Up till now the parameter/state variable has been a fixed, but unknown value. In this section we shall introduce a different philosophy, where the parameter is assumed to be unknown and have a distribution, and we will make active use of it. The parameter distribution may be subjectively or objectively based, but in any case this distribution will change when we make an observation. An observation will bring us more information about the parameter, and in general this will narrow its distribution. We talk about a *prior* distribution before the experiment and a *posterior* distribution after the experiment.

To study this, we first develop the simplest case of *Bayes' formula*, which is essentially just a refinement of the definition (1.2) of a conditional probability.

Theorem 2.6.1. *Let B be some fixed event, and let $A_1, A_2, \ldots, A_k$ be a partition of the probability space, i.e., disjoint event such that $A_1 \cup A_2 \cup \ldots \cup A_k = \Omega$ and $P(A_i) > 0$; $i = 1, \ldots, k$. Then*

$$P(A_i|B) = \frac{P(A_i \cap B)}{P(B)} = \frac{P(A_i)P(B|A_i)}{\sum_j P(A_j)P(B|A_j)}. \tag{2.33}$$

Here $P(A_i)$ can be considered as a prior probability, while $P(A_i|B)$ is the posterior probability after the observation B is done. The statistical model is given by $P(B|A_i)$. So this is an exact formula for updating from the prior to the posterior depending only upon the model for observations.

Similarly, in the case of continuous observations y depending upon continuous parameters θ through a model with density $q^\theta(y)$ and with a prior $\nu(\theta)d\theta$, Bayes' formula reads

$$\nu(\theta|y) = \frac{\nu(\theta)q^\theta(y)}{\int_{\theta'} \nu(\theta')q^{\theta'}(y)d\theta'}. \tag{2.34}$$

Here $\nu(\theta|y)$ is the posterior density of θ.

This in principle gives a completely new theory of statistical inference compared to what has been discussed above: In (2.34) we have obtained a direct probability distribution of the unknown parameter depending upon the experimental data. However, there are two obstacles to these Bayesian methods in practice: First you have to provide a reasonable prior distribution. Next, finding the posterior involves quite complicated integration in most cases, integration that usually will have to be carried out in some numerical way. In recent years, Markov Chain Monte Carlo methods [91] have made these integration problems feasible, and this has lead to an upswing in the use of Bayesian methodology.

As a first item, we define the concept of *credibility interval*.

Definition 2.6.1. *For a scalar parameter θ a credibility interval is an interval $[c_1(y), c_2(y)]$ such that, say*

$$P(c_1(y) \leq \theta \leq c_2(y)) = \int_{c_1(y)}^{c_2(y)} \nu(\theta|y)d\theta = 0.95. \qquad (2.35)$$

The form of the Bayesian credibility interval is like that of the confidence interval, but the interpretation is completely different: For fixed data, the probability that the parameter is in the interval is equal to the credibility coefficient, here 0.95. Surprisingly, however, it will turn out later that in some natural cases the confidence interval and the credibility interval with the same coefficient will in fact be numerically equal.

A related question is how to estimate a parameter θ. We then need to specify a *loss function*.

Definition 2.6.2. *A loss function L is a function of two variables such that $L(\hat{\theta}, \theta)$ gives the loss you experience when the real value is θ and you estimate it by $\hat{\theta}$: If the distance is large, you suffer a large loss. and nearly no loss if the distance is small.*

From this you calculate the *risk function* or expected loss:

Definition 2.6.3. *The risk function is given by*

$$R(\theta) = E^\theta L(\hat{\theta}, \theta) = \int_y L(\hat{\theta}(y), \theta)q^\theta(y)dy. \qquad (2.36)$$

The risk function depends upon the unknown parameter θ, and it is much used in traditional (frequentist; cp. Section 2.2–2.5 above) inference theory, too.

In Bayesian theory, in principle at least, everything is easier: We have the possibility of integrating out the unknown parameter by using the posterior. This gives the *Bayes risk*:

Definition 2.6.4. *The Bayes risk is given by*

$$r = E^\nu R(\theta) = \int_\theta R(\theta)\nu(\theta|y)d\theta. \qquad (2.37)$$

This can be calculated for several proposals for estimator, so that we can choose the one which gives the smallest r.

For scalar θ, a very common choice of loss function is the quadratic loss

$$L(\hat{\theta}, \theta) = (\hat{\theta} - \theta)^2, \tag{2.38}$$

which gives a risk function equal to the mean square error (2.3)

$$R(\theta) = ME = (E(\hat{\theta}) - \theta)^2 + Var(\hat{\theta}) = E^\theta(\hat{\theta} - \theta)^2.$$

A mathematical exercise ([24], p. 161) shows:

Lemma 2.6.1. *For quadratic loss the Bayes risk is minimized by taking the posterior mean as an estimator:*

$$\hat{\theta} = \hat{\theta}(y) = E(\theta|y) = \int_\theta \theta \nu(\theta|y) d\theta. \tag{2.39}$$

Note that this indeed is an estimator, since it only depends upon the observations y.

Example 2.6.1. ([24]) Assume one observation y which is normal (θ, σ^2), where θ is unknown but σ is known. Let the prior for θ have a normal (μ, τ^2) density $\nu(\theta)$, where both μ and τ are known. Then

$$\nu(\theta)q^\theta(y) = (2\pi\sigma\tau)^{-1}\exp\{-\frac{1}{2}[\frac{(\theta - \mu)^2}{\tau^2} + \frac{(y - \theta)^2}{\sigma^2}]\}.$$

By using $\rho = \tau^{-2} + \sigma^{-2}$ this can be transformed by completing squares to

$$(2\pi\sigma\tau)^{-1}\exp\{-\frac{1}{2}\rho[\theta - \frac{1}{\rho}(\frac{\mu}{\tau^2} + \frac{y}{\sigma^2})]^2\}\exp\{-\frac{(\mu - y)^2}{2(\sigma^2 + \tau^2)}\},$$

and the integral of this over θ is

$$(2\pi\rho)^{-1/2}(\sigma\tau)^{-1}\exp\{-\frac{(\mu - y)^2}{2(\sigma^2 + \tau^2)}\}.$$

The posterior is a normal density with expectation $E(\theta|y)$, and with variance ρ^{-1}. Then, from the last two equations we read out that the Bayes estimate is

$$\hat{\theta} = E(\theta|y) = \frac{1}{\rho}(\frac{\mu}{\tau^2} + \frac{y}{\sigma^2}) = \frac{\sigma^2}{\sigma^2 + \tau^2}\mu + \frac{\tau^2}{\sigma^2 + \tau^2}y. \tag{2.40}$$

This last expression has a very simple interpretation: The estimator is a weighted mean between the prior value μ and the data value y. When the prior variance τ^2 is large compared to the data variance σ^2, the largest weight is on the data and vice versa.

Consider now the case of a series $y = y_1, \ldots, y_n$ of independent normal (θ, σ) measurements. Then $\bar{y}$ is normal $(\theta, \sigma^2/n)$, and using just this observation (cp.

sufficiency; see Section 2.11 below), everything is reduced to the situation of the previous paragraph. Thus

$$\hat{\theta} = E(\theta|y) = \frac{\sigma^2}{\sigma^2 + n\tau^2}\mu + \frac{n\tau^2}{\sigma^2 + n\tau^2}\bar{y}. \tag{2.41}$$

The interpretation is as before with the addition that a large value of n implies that one has a large weight on the data value $\bar{y}$.

As this example shows, the calculation involved in Bayesian inference can be quite cumbersome even in the simplest cases.

Example 2.6.2. Look again at the uniform distribution (1.11), and as in Example 2.3.2 and Example 2.3.6 we seek estimators for the parameters α and β. To find the Bayesian estimators (2.39), we first choose a uniform improper joint prior for the parameters, and then from this find the aposteriori distribution by Bayes formula. Here the prior is improper, but this does not matter as long as the posterior is a proper distribution. The joint aposteriori density for $\alpha \le \min(y_i), \beta \ge \max(y_i)$ is

$$\nu(\alpha, \beta|y) = \frac{(\beta - \alpha)^{-n}}{\int_{\alpha=-\infty}^{\min(y_i)} \int_{\beta=\max(y_i)}^{\infty} (\beta - \alpha)^{-n} d\alpha d\beta}$$

$$= \frac{(n-1)(n-2)(\max(y_i) - \min(y_i))^{n-2}}{(\beta - \alpha)^n}.$$

Integrating out β gives

$$\nu(\alpha|y) = \frac{(n-2)(\max(y_i) - \min(y_i))^{n-2}}{(\max(y_i) - \alpha)^{n-1}}$$

for $\alpha \le \min(y_i)$, and from this

$$\hat{\alpha} = \int \alpha\nu(\alpha|y)d\alpha = \min(y_i) - \frac{\max(y_i) - \min(y_i)}{n-3}.$$

By symmetry

$$\hat{\beta} = \max(y_i) + \frac{\max(y_i) - \min(y_i)}{n-3}.$$

These estimators are quite close to the maximum likelihood estimators, and seem to have good properties.

The Bayesian analysis and Bayesian estimation will in Chapter 3 be associated with a transformation group on the parameter space. The most natural transformation group here seems to consists of scale transformations in $\delta = \beta - \alpha$ combined with translations in $\mu = (\alpha + \beta)/2$, which implies a different prior than the one that was used above. For the Bayes estimator obtained from the latter prior, see Example 3.5.1.

In the statistical community there has been a discussion between the classical statisticians and the Bayesian for decades. The distance between the two schools have diminished in the last decades, however. A mathematical reason for this ([145]) is a basic result of Wald to the effect that every reasonable classical decision procedure is a Bayes solution or a limit of Bayes solutions. Another reason is that many statisticians tend to be pragmatic and use the method that they judge to be best on each given problem. If you have a reasonable prior, are able to carry out the integrations and think that the problem at hand is worth the effort, you use a Bayesian method, otherwise the classical methodology in most cases offers plenty of possibilities. Note that the choice of a statistical procedure is not like the solution of a mathematical problem, even though some students seem to believe that, and indeed sometimes are given an education which encourages them to believe that. In practice there is usually not one single best procedure, but the analysis can be carried out in several reasonable ways. This of course does not prevent us in believing that there is an objective world behind our analysis of uncertain data.

2.7 Regression and Analysis of Variance

Regression analysis and analysis of variance belong to the most used methods in statistical practice. They are tied together by the class of linear models, a class which in our setting also will be used below to illustrate the diversity in applied science concerning the parameter concept. Usually these models are analysed in a frequentist way, so we will concentrate on this, but there exist also Bayesian variants.

Example 2.7.1. In Example 2.5.2 we indicated how to test two diets for slimming purpose against each other. Assume now that we want to compare k different diets. Select randomly n_j persons for diet j $(j = 1, \ldots, k)$ and let the observed weight losses be $\{y_{jr}; j = 1, \ldots, k; r = 1, \ldots, n_j\}$. This is called a one way analysis of variance situation, and the natural model is that the observations y_{jr} are independent and normally distributed (μ_j, σ). Equivalently, we may write

$$y_{jr} = \mu_j + e_{jr}, ; \; j = 1, \ldots, k; \; r = 1, \ldots, n_j, \qquad (2.42)$$

where the error terms e_{jr} are independent and normal $(0, \sigma)$. The full parameter is $\theta = (\mu_1, \ldots, \mu_k, \sigma)$.

Example 2.7.2. There can be given many examples of data pairs $\{(x_i, y_i)\}$ where it is natural to try a simple linear fit. Such examples can be found in almost any textbook. Most often, the x_i's can be taken as values of some explanatory, fixed variable, and the y_i's as some random response. The model usually posed for such data is

$$y_i = \beta_0 + \beta_1 x_i + e_i; \; i = 1, \ldots, n, \qquad (2.43)$$

where again the error terms e_i are independent and normal $(0, \sigma)$. The full parameter is $\theta = (\beta_0, \beta_1, \sigma)$.

Example 2.7.3. A full IQ test can be fairly complicated. Imagine that we want to predict IQ from a series of p simple tests, perhaps each of them consisting of just a single test question. This is investigated further for n randomly selected persons, for which one on person i measures the IQ score y_i and the score x_{ij} on the simple test number j $(j = 1, \ldots, p))$. Here we want to study the effect of x upon y, so it is natural to condition upon the x-variables. This means that one also here takes the x_{ij}'s as fixed. The model most often used in such situations is

$$y_i = \beta_0 + \beta_1 x_{i1} + \ldots + \beta_p x_{ip} + e_i; \ i = 1, \ldots, n, \tag{2.44}$$

where once more the error terms e_i are independent and normal $(0, \sigma)$. The full parameter is $\theta = (\beta_0, \beta_1, \ldots, \beta_p, \sigma)$.

It is obvious that model (2.43) is a special case of model (2.44). But also model (2.42) can be taken as a special case of this model if the y_{jr}'s are stacked, so that (jr) is equivalent to a new index i, and the x-variables are indicators of the diet chosen, specifically $x_{ij} = 1$ if person i has taken the diet j, otherwise $x_{ij} = 0$. Then take $\beta_j = \mu_j$; $j = 1, \ldots, k$ and $\beta_0 = 0$.

The model (2.44) is called the general linear model, and it has these 3 applications and a lot of other applications. On vector form the model equation is written

$$y = X\beta + e, \tag{2.45}$$

with obvious definitions of the n-vectors y and e, the $n \times (p + 1)$-matrix X and the $(p + 1)$-vector β. It is convenient to write $E(y)$ for the vector of expectations $E(y_i)$ and $V(y)$ for the covariance matrix of y, i.e., the matrix with variances on the diagonal and covariances $Cov(y_i, y_j) = E(y_i - E(y_i))(y_j - E(y_j))$ on the off-diagonal entrices. With the model assumptions above, we have simply

$$V(y) = V(e) = \sigma^2 I, \tag{2.46}$$

where I is the identity matrix.

The regression vector estimates are developed in Appendix A.1.4. They are found by least squares, in this case the same as the maximum likelihood method, i.e. by minimizing $\| y - X\beta \|^2$. The solution is

$$\hat{\beta} = (X'X)^{-1}X'y, \tag{2.47}$$

which gives

$$E(\hat{\beta}) = \beta, \quad V(\hat{\beta}) = (X'X)^{-1}\sigma^2. \tag{2.48}$$

The last equation is interesting. If $X'X$ is singular, which occurs if X has rank less than $p + 1$, then then the covariance matrix is infinite, and there is at least one linear combination of the β's which does not have a value in the experiment. If $X'X$ is nearly singular, then the estimates are unstable. In such cases a model reduction is often called for.

An unbiased estimate of σ^2 is

$$s^2 = \frac{\parallel y - X\hat{\beta} \parallel^2}{n - p - 1}. \tag{2.49}$$

From this, a standard error $se(\hat{\beta}_j)$ is found for each regression parameter estimator by replacing σ by s in the last part of (2.48). This gives single parameter confidence intervals $\hat{\beta}_j \pm t_0 se(\hat{\beta}_j)$, where t_0 is found from the t-table with $n - p - 1$ degrees of freedom. Similarly, single parameter t-tests are found.

A test for the joint null hypothesis $\beta_1 = \beta_2 = \ldots = \beta_p = 0$ can be constructed in the following way: Calculate first the total sum of squares $SST = (n-1)s_y^2 = \sum(y_i - \bar{y})^2$, then the regression sum of squares from $SSR = SST - SSE$, where $SSE = (n - p - 1)s^2 = \parallel y - X\hat{\beta} \parallel^2$, and then finally the test variable

$$F = \frac{SSR}{ps^2}. \tag{2.50}$$

Then the null hypothesis is rejected if $F > f_0$, a one-sided test, where f_0 is found from the F-distribution of Definition 2.5.3 with p and $n - p - 1$ degrees of freedom. This test is often done as a first step in the investigation of a regression model. If the null hypothesis is not rejected here, there is usually little reason to continue the investigation.

By using a similar procedure, one can test hypotheses of the form $\beta_{q+1} = \ldots = \beta_p = 0$ and even more general hypotheses.

All this is standard theory which can be found in many textbooks, for instance Weisberg [201], where many applications are found. The results above are proved in an analogous way to those of Section 2.5 by using matrix calculations and as indicated in Appendix A.1.4.

All the procedures above assume that the model used is true, an issue which is looked at in the next section.

Example 2.7.4. Consider again the situation of Example 2.7.3, but now with the following modification: Either in order to avoid problems with estimation or because the capacity of each subject is limited, we feel forced to drop some of the simple tests from the investigation. Specifically, it is determined that only q of the p tests shall be included in the experiments, but the choice of q tests among the p is left open.

This leaves one with the choice between $\binom{p}{q}$ experiments. Experiment a corresponds to a certain selection of indices $j_1, \ldots, j_q$, and the full parameter for this experiment is $\theta^a = (\beta_0, \beta_{j_1}, \ldots, \beta_{j_q}, \sigma)$.

This simple example has in fact many features in common with the setting that we shall assume later for quantum mechanics. The different parameters θ^a are complementary. The parameter θ^a must be assumed relative to experiment a to take some value which can be estimated accurately in a perfect experiment, say with a large number of subjects n. But relative to another experiment b, the full parameter θ^a can not be assumed to take a value. However, some components may have values in the two different experiments; these may or may not be equal (there may be interactions between the tests).

There is a c-variable $\phi = (\beta_0, \beta_1, \ldots, \beta_p, \sigma)$ so that every θ^a is a function of ϕ. This ϕ will not take a value in the setting above, but nevertheless, symmetry properties may be defined relative to ϕ. In later chapters we will meet several cases like this, where an estimable parameter is a function of a total, inaccessible c-variable.

Given strong enough symmetry properties, one might hope to be able to predict probabilistically θ^b from θ^a. In fact, in Chapter 5, I will describe quantummechanical methods for similar predictions from one experiment to another under symmetry assumptions.

It must be emphasized, however, that I do do not intend to say that the simple example above mimics exactly the properties of quantum objects. Our intentions with such discussions around certain examples in this and the previous chapter is to illustrate how flexible the parameter/ c-variable concept must be in classical statistics, and then come back in Chapters 4, 5 and 6 and argue that quantum mechanics can be derived and discussed under the same umbrella. Further such discussions of classical statistics related to the linear model will be given in the following sections.

2.8   Model Checking in Regression

In the previous section we took the linear model (2.44)-(2.45) as a point of departure. In fact, in the last decades there has been developed many tools to check such models. If it is decided that the model is wrong in some way, either data must be further investigated and perhaps deleted, or the model must be changed, possibly by adding new parameters.

According to a definition sometimes employed in standard statistics, the data together with the original model is regarded as an 'experiment' with some (full) parameter θ. If this view is followed, additional parameters found from model checks must be regarded as complementary parameters. In contrast to the situation in Example 2.7.4, they can be estimated by using the same data, however.

Specifically, for a linear model, most model checks are made by looking at *residuals*. The original model is used for estimation, and the n-dimensional residual vector is given as

$$\hat{e} = y - X\hat{\beta}. \tag{2.51}$$

The residuals $\hat{e}_i$ are plotted against time, against potential new x-variables, against the original variables x_{ij} for each fixed j or against the predicted values $\hat{y}_i = (X\hat{\beta})_i$. Here are some very common situations:

(a) $\hat{e}_i$ plotted against $\hat{y}_i$ shows some outliers. These must be removed or modelled by a new parameter θ^a.

(b) $\hat{e}_i$ plotted against x_{ij} shows that the spread of the residuals is not constant, but increases as x_{ij} increases. A transformation of y might be possible in this case, but otherwise a new model with σ^2 is taken as a function of x_{ij} through some parameter θ^b might be the solution.

(c) $\hat{e}_i$ plotted against x_{ij} shows some curvature. In a new model a second degree term with a coefficient θ^c is added.

(d) $\hat{e}_i$ plotted against a new variable x_{i0} shows that there is a positive or negative slope. A new model with a term $\theta^d x_{i0}$ might be called for.

2.9   Factorial Models

The one way model illustrated in Example 2.7.1 is only the simplest of a large class of analysis of variance models. Each of these can be written as linear models with a design matrix X composed of 0's and 1's.

Example 2.9.1. Assume that one wants to study weight reduction not only across diets, but also across age groups (and perhaps also across sexes). Then one has to perform a *factorial* experiment: Persons have to be sampled for each combination of diet and age group (and perhaps also sex).

Assume now in general a factorial experiment with two factors. Let there be k levels (number of possible values) of the first factor (say diet) and m levels if the second factor (say age group) Let y_{jlr}; $j = 1, \ldots, k$; $l = 1, \ldots, m$; $r = 1, \ldots, n_{jl}$ be the responses (say weight decrease) for the persons at each combination of factorial levels. In the usual model these are assumed to be independent and normal with constant variance σ^2 and expectation

$$E(y_{jlr}) = \mu_{jl} = \mu + \alpha_j + \beta_l + \gamma_{jl}. \tag{2.52}$$

The last parametrisation is very useful in the interpretation of results: α_j is the main effect of diet number j, say, β_l is the main effect of age group l, while γ_{jl} is the socalled interaction. In a model without interaction there is simple additivity between the two factors, but not when interaction is present.

Let us first look at the balanced case where n_{jl} is constant. In an optimal design this should indeed be the case. Then it is common to use the constraints

$$\sum_j \alpha_j = \alpha. = 0; \quad \beta. = 0; \quad \gamma.. = 0. \tag{2.53}$$

In fact, these are automatic if one defines

$$\mu = \bar{\mu}.. = \mu../km; \quad \alpha_j = \bar{\mu}_j. - \mu; \quad \beta_l = \bar{\mu}._l - \mu; \quad \gamma_{jl} = \mu_{jl} - \bar{\mu}_j. - \bar{\mu}._l + \mu. \tag{2.54}$$

Here we have used the dot-notation for sum over the levels of one factor. Means are denoted as in $\bar{\mu}_{..}$ above, and $\bar{\mu}_{.l} = k^{-1} \sum_{j=1}^{k} \mu_{jl}$.

In this case all parameters are uniquely defined, and can easily be estimated. Confidence intervals and tests can be developed from ordinary linear models theory.

If the estimated interaction is small, it can often be argued that one should use the simpler model obtained by putting $\gamma_{jl} \equiv 0$. Such model reductions are common in applied statistics, and will be further discussed later. But note already now that not all model reductions are meaningful. A model with $\alpha_j = 0$ and $\gamma_{jk} \neq 0$ is very awkward. Such problems were taken up broadly by Nelder [160], and we will return to them in Chapter 3.

Turn now to the unbalanced case, i.e., where the number of observations n_{jlr} is not constant. Typically, an investigation may be designed to be balanced, but there may be missing observations. For this case it is argued by Searle [183] that one should not use any constraints. The model should be taken as in (2.52), but the individual parameter, say, α_j should not be taken to have a value. As described in detail in [183], the estimatian could be carried out by using an arbitrary generalized inverse instead of $(X'X)^{-1}$, but this is then non-unique. However, certain linear combinations of the parameters are *estimable* and have unique estimates. These turn out to be all parameters which are linear combinations of the $E(y_{jlr})$, for instance $E(\bar{y}_{jl.}) = \mu + \alpha_j + \beta_l + \gamma_{jl}$ or $E(\bar{y}_{1..}) - E(\bar{y}_{2..}) = \alpha_1 - \alpha_2 + \bar{\gamma}_{1.} - \bar{\gamma}_{2.}$. Similarly, certain hypotheses are testable, and are the hypotheses of interest. The analysis of unbalanced model becomes quite cumbersome this way, but the analysis makes sense also for instance in the case of missing data. The computer package SAS analyses unbalanced data essentially in this way.

For our purpose it is useful to observe that statistical models can make sense also for models with parameters which do not take a value relative to the experiment put up. If this is the only experiment for which the relevant parameter can be estimated, this comes close to our general notion of not taking a value, compare also the related property of identifiability in Definition 1.10.1.

2.10   Contrasts in ANOVA Models

We now return for simplicity to the one way analysis of variance case, say illustrated by the diet example 2.7.1. A first step is an overall test for the difference between the diets, but often we want more information. Typically diets will have some structure relative to each other, say the first two could be lowfat diets and the last 3 are low carbohydrate diets, and such structure is made use of in the statistical analysis.

For a one way analysis model with k treatments (diets) we parametrize the model as

$$E(y_{jr}) = \mu_j = \mu + \alpha_j. \tag{2.55}$$

The α_j's can be constrained or unconstrained; it does not matter in this case.

The first step is to test $\alpha_1 = \ldots = \alpha_k = 0$. It is a special case of the test given in connection to (2.50), and it has $k-1$ degrees of freedom in the numerator. This correponds to a possibility of investigating $k-1$ comparisons among the treatments.

Example 2.10.1. Consider the diet example. Assume that k=4 diets are included in the investigation: FC, fC, Fc and fc, where f is low fat, F is high fat, C is high carbohydrate and c is low carbohydrate. Assume also that one primarily is interested in finding out about the effect of carbohydrate on the response. Then one first puts up the *contrast*

$$\tau_1 = \frac{1}{2}(\mu_{FC} + \mu_{fC}) - \frac{1}{2}(\mu_{Fc} + \mu_{fc}),$$

which compares the two levels of carbohydrate. In addition one can be interested in comparing the effect of fat within each of the carbohydrate levels. This gives two new contrasts

$$\tau_2 = \mu_{FC} - \mu_{fC}$$

and

$$\tau_3 = \mu_{Fc} - \mu_{fc}.$$

Note that these contrasts are orthogonal: The product-sum of the coefficients for each pair is zero. In the balanced case this implies that the corresponding estimators are uncorrelated, which in a normal model means that they are independent.

In all this gives $k-1 = 3$ comparisons, and this is the maximal number one can have with $k = 4$ treatments. The dimension of the parameter space is 4, but one dimension (called degree of freedom) is used to estimate the overall mean μ.

Definition 2.10.1. *In general, a contrast is a linear combination*

$$\tau = \sum_j c_j \mu_j = \sum_j c_j \alpha_j \tag{2.56}$$

such that $\sum_j c_j = 0$. *Two contrasts with coefficients* $\{c_j\}$ *and* $\{d_j\}$ *are called orthogonal if* $\sum_j c_j d_j = 0$.

As in Example 2.10.1, orthogonal contrasts imply in the balanced case independent estimators $\sum_j c_j \bar{y}_{j\cdot}$.

For a given contrast τ one can find a confidence interval from the estimator, and there is a t-test for the hypothesis $H_\tau : \tau = 0$. Since the estimators are independent for orthogonal contrasts, the tests here can be considered to concern orthogonal, unrelated questions.

Any contrast and any orthogonal set of contrasts can be extended to a maximal set of $k-1$ orthogonal contrasts. Together with the total mean $\mu = \bar{\mu}$. this set of contrasts is in one-to-one correspondence with $(\mu_1, \ldots, \mu_k)$. Thus the hypothesis that all contrasts are zero can be tested with the clearance test mentioned above. The set of orthogonal contrasts can in general be chosen in many complementary ways.

Example 2.10.2. Return to Example 2.10.1, but assume instead that one primarily is interested in effect of the fat levels. Then it is natural to look at the orthogonal contrasts

$$\tau_1' = \frac{1}{2}(\mu_{FC} + \mu_{Fc}) - \frac{1}{2}(\mu_{fC} + \mu_{fc}),$$

$$\tau_2' = \mu_{FC} - \mu_{Fc},$$

$$\tau_3' = \mu_{fC} - \mu_{fc}.$$

Such sets of investigations must be said to be complementary even though the same experimental data are used and the sets of contrasts are functions of each other. Complementarity of sets of contrasts in a stronger sense can occur in connection to socalled incomplete block design [154], where the number of treatments is larger than the size of the homogeneous blocks into which the experimental units are sorted in order to enhance efficiency, but this is beyond the scope of this book.

2.11   Reduction of Data in Experiments: Sufficiency

We now turn again to the case of one standard statistical experiment with some given model. We can learn something, however, from the examples discussed above: In the normal measurement series case we do not need all the observations to do a statistical analysis, it is enough to know $t = (\bar{y}, s^2)$. In the linear model it can be shown that every analysis can be carried out if we know the estimated regression coefficients and s^2.

In general, then, consider a standard statistical experiment case with a model Q^θ.

Definition 2.11.1. *Let the complete data be y, and let us define a statistic as any function $t(y)$ of the data. Such a statistic is called* sufficient *if the conditional distribution of y given t does not depend upon θ.*

This means in effect that all information about the parameter θ is contained in t.

Lemma 2.11.1. *If y has a probability density $q^\theta(x)$ and t has a probability density $g^\theta(u)$, then $t = t(y)$ is sufficient if and only if $q^\theta(x) = k(x)g^\theta(t(x))$ for some function θ-independent function $k(x)$.*

A sufficient statistic is *minimal* if it is a function of any other sufficient statistic. For example, the full data y is trivially always sufficient, but this is not of much interest. A good reduction $t(y)$, like in the normal measurement series case, see Example 2.3.1, is much more interesting, especially if it is minimal. There exist cases where a minimal sufficient statistic does not exist, but we have the following result:

Lemma 2.11.2. *Let the density of y be of exponential family form:*

$$q^\theta(x) = k(x)p(\theta)\exp(c(\theta)t(x)), \tag{2.57}$$

where $c(\theta)$ and $t(x)$ are vectors of the same dimension and $c(\theta)t(x)$ is understood as a scalar product. If the components of $c(\theta)$ are linearly independent over Θ, the space of parameter values θ, then $t(y)$ is minimal sufficient.

The form (2.57) may appear special, but it contains many important particular cases, for instance the multivariate normal distribution. It is a straightforward exercise to prove from this criterion that $t = (\bar{y}, s^2)$ is sufficient in a normal measurement series, in fact, it is minimal sufficient.

So, how can one state in some precise way that it is useful to reduce data to a (minimal) sufficient statistic? One way is as follows: Let $\hat{\theta} = \hat{\theta}(y)$ be any estimator of θ. Then consider $\hat{\theta}_0 = E(\hat{\theta}|t)$. This is independent of θ, so it is an estimator. It is unbiased if $\hat{\theta}$ is so. Now the well-known Rao-Blackwell theorem (see for instance [146]) says:

Theorem 2.11.1. *If one has a convex loss function, then the risk function (cp. section 2.6 above) for $\hat{\theta}_0$ is always smaller or equal than for $\hat{\theta}$. Thus the estimator $\hat{\theta}_0$ depending upon the minimal sufficient statistic is always the best one.*

Definition 2.11.2. *A sufficient statistic t is called* complete *if for any integrable $h(t)$ we have that*

$$E^\theta(h(t)) = 0 \text{ for all } \theta \text{ implies } P^\theta(h(t) = 0) = 1 \text{ for all } \theta. \tag{2.58}$$

A complete sufficient statistic is minimal sufficient (a proof of this is given in [147]), but a minimal sufficient statistic needs not be complete.

There is a criterion for complete statistics related to exponential models, too:

Lemma 2.11.3. *If the model (2.57) holds and $c(\Theta) = \{c(\theta); \theta \in \Theta\}$ contains an open set, then $t(y)$ is a complete sufficient statistic.*

All these results, and more results, examples and references can be found in textbooks like [146] or in the encyclopedia article [12].

2.12   Fisher Information and the Cramér-Rao Inequality

Consider again a standard statistical experiment. We have earlier defined the likelihood function $L(\theta) = q^\theta(y)$ and the log likelihood $l(\theta) = \ln L(\theta)$. The maximum likelihood estimation method aims at maximizing L or l.

Definition 2.12.1. *Define the score function $s(\theta)$ by*

$$s(\theta) = \frac{\partial}{\partial \theta} l(\theta). \tag{2.59}$$

The score function is needed, say, when using Newton's method to find the maximum likelihood numerically. Like $l(\theta)$, the score function $s(\theta)$ is a random variable for each fixed θ, and in Appendix A.1.5 it is shown that $E^\theta(s(\theta)) = 0$.

Definition 2.12.2. *We define the* Fisher information *by*

$$I(\theta) = Var^\theta(s(\theta)). \tag{2.60}$$

In Appendix A.1.5, the Cramér-Rao inequality is proven:

Theorem 2.12.1. *For every unbiased estimator $\hat{\theta}$ of θ we have*

$$Var^\theta(\hat{\theta}) \geq \frac{1}{I(\theta)}. \tag{2.61}$$

Under regularity conditions one can show [146] that the maximum likelihood estimator for a series of n measurements is asymptotically optimal in the sense that $Var^\theta(\hat{\theta}) \approx \frac{1}{I(\theta)} \approx \frac{1}{I(\theta)}|_{\theta=\hat{\theta}}$ for large n.

The Cramér-Rao inequality, which is useful in several areas of theoretical statistics can alternatively be written

$$Var^\theta(\hat{\theta}) Var^\theta(s(\theta)) \geq 1, \tag{2.62}$$

and it thus has some structural similarity with the Heisenberg inequality (Theorem 1.1.1) for position and velocity. Williams [207] attributes this to the fact that the Cauchy-Schwarz inequality has been used in the proof both places.

A deeper connection is pointed out by a referee: One natural connection is by using commutativity. The operators D and E given by $Df = \partial f/\partial \theta$ and $Ef = \theta f$ have a commutator $DE - ED$ equal to the identity I, compared to the commutator $\hbar I$ between the momentum and position operator in the Heisenberg case. I will come back to a more concrete discussion of this connection in Chapter 7.

2.13 The Conditionality Principle

I have above given several examples of a *standard statistical experiment*, i.e., an experiment given by a parameter space Θ, a sample space of observations and a model Q^θ which is a probability distribution of the observations depending upon the parameter θ. It must be noted, however, that it was illustrated in several examples above that one sometimes in practical experiments may have to go beyond this simple frame.

This section concerns sets of hypothetical experiments; thus it is related to the discussion of counterfactuals in Section 1.6.

Example 2.13.1. (Cox [48])
Consider two potential laboratory experiments for the same unknown parameter θ such that $\mathcal{E}^1$ is planned to be carried out in New York while $\mathcal{E}^2$ is planned to be carried out in San Francisco. The owner of the material that shall be sent to these laboratories chooses to toss an unbiased coin in order to decide the laboratory. So $\mathcal{E}^1$ will be chosen with probability $1/2$ and $\mathcal{E}^2$ will be chosen with probability $1/2$.

Now consider the whole experiment $\mathcal{E}$ including the coin toss. If $\mathcal{E}^1$ and $\mathcal{E}^2$ are standard statistical experiments, then $\mathcal{E}$ is also a standard statistical experiment. The fundamental question is then: Does $\mathcal{E}$ in any sense contain more information about the parameter θ than the chosen experiment $\mathcal{E}^i$?

No, says common sense, and also Cox [48]: Once the chosen experiment has been carried out, it is completely irrelevant that it was chosen by some coin toss mechanism.

This example has been turned out into a general principle.

The conditionality principle. *Let $\mathcal{E}^i$ $(i = 1, \ldots, k)$ be k experiments, and let y^i $(i = 1, \ldots, k)$ be the outcomes of these experiments. Let π^i $(i = 1, \ldots, k)$ be k fixed probabilities with $\pi^1 + \ldots + \pi^k = 1$, and define a new experiment $\mathcal{E}$ as follows: Choose $\mathcal{E}^i$ $(i = 1, \ldots, k)$ with probability π^i $(i = 1, \ldots, k)$ and observe the outcome (i, y^i). Then the experimental evidence obtained by $\mathcal{E}$ is equal to that of the chosen experiment:*

$$\mathrm{Ev}\{\mathcal{E}, (i, y^i)\} = \mathrm{Ev}\{\mathcal{E}^i, y^i\}. \tag{2.63}$$

The central concept 'experimental evidence' is left undefined; it could be anything related to the unknown parameter θ.

This principle is closely related to a normative conditionality principle: All inference should be made conditionally, given an ancillary statistic (a statistic whose distribution is independent of the parameters, like the result of the coin toss above.) This claim goes back to R.A. Fisher's writing, and was discussed and related to

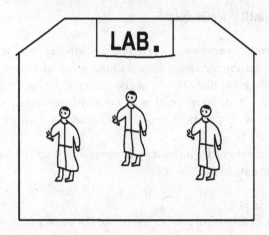

Fig. 2.1   Laboratory with three assistants.

concrete examples in [48]. In the formulation of the conditionality principle, the variable i plays the rôle of the ancillary statistic.

In fact, it is a problem that cases with several ancillaries exist such that the joint statistic is not an ancillary; this is a problem that we will come back to at the end of the section.

The question we will address first is if there are other limitations to the seemingly obvious conditionality principle. The following example is a bordering case.

Example 2.13.2.

Consider a laboratory where chemical samples of a certain type are analysed routinely. There are three laboratory assistants having the same training, using similar equipment etc.. For each sample that arrives for analyis the administrator of the laboratory selects randomly two of the three assistants; these two make independent analyses and their average result is reported. From the point of view of the momentary user of the laboratory, 3 different experiments can be identified: $\mathcal{E}^{12}$, $\mathcal{E}^{13}$ and $\mathcal{E}^{23}$, according to which pair among the laboratory assistants 1, 2 and 3 that is selected for the job. The momentary user may perhaps want to condition upon the selected experiment to assess the accuracy of the analysis that has been done, particularly if he knows something about the particular laboratory assistants. If such knowledge is missing, however, it is not possible to achieve anything by conditioning. From the point of view of the laboratory, the random selection of assistants is a part of their routine, and the natural thing to do is to report routinely an unconditional standard deviation connected to their result.

Note that this example in reality goes beyond the frame of one single standard statistical experiment. In the unconditional report from the laboratory a different model must be imagined to have been used than the detailed model with specific information about each pair of laboratory assistants.

A similar discussion can be made in a sensory analysis if there is a staff of N trained assessors and a panel of n out of these are selected randomly to taste a particular product. Factors that may be of importance in favour of reporting an unconditional error are: The sensory analysis firm with the N assessors may constitute a unit; the testings are performed routinely; the random selection of assessors may be a part of the procedure; the number n is not too small.

The following simplified example from Helland [107] may throw some light upon this general discussion:

Example 2.13.3.

As a small part of a larger medical experiment, two individuals (1 and 2) have been on a certain diet for some time, and by taking samples at the beginning and at the end of that period some response like the change in blood cholesterol levels is measured. For the individual i $(i = 1, 2)$, the measured response is y_i, which we suppose is normal (μ_i, σ^2) with a known measurement variance σ^2. Measurements on different individuals are independent.

Because these two individuals have been given the same treatment (diet) in the larger experiment, the parameter of interest is not μ_1 or μ_2, but their mean: $\tau = \frac{1}{2}(\mu_1 + \mu_2)$.

The experiments $\mathcal{E}^1$ and $\mathcal{E}^2$ are defined defined simply as measuring y_1 and y_2, respectively. It is crucial to make precise what we mean by experimental evidence here, namely to give some information about τ, the average expected value.

It is clear that neither $\mathcal{E}^1$ nor $\mathcal{E}^2$ can give any such information. Now let $\mathcal{E}$ be the the experiment constructed by selecting $\mathcal{E}^1$ with probability $1/2$ or $\mathcal{E}^2$ with probability $1/2$. I claim that this experiment contains some nontrivial information in the sense just described. Namely: We choose here randomly 1 out of 2. What if we instead had chosen n out of N? Then this would have been a simple sampling experiment. It is difficult to claim that $n = 1; N = 2$ is different in principle.

In a discussion of this example, Aitkin [10] pointed out that it is not a couterexample to the conditionality principle: In this principle, it is assumed that i in the experiment $\mathcal{E}^i$ is known, while in the sampling experiment i is unknown. In the example, if it is known that $i = 1$, for instance, it is known that the parameter in question is μ_1, so the example gives no information about τ.

Nevertheless, the example is of interest. It points at the difference between knowing i and not knowing i, and similarly in a larger sampling experiment. In this case, we can make more inference when we have less knowledge!

Consider first this variant. None of the individuals is known to the experimentalist; they are just identified as numbers 1, 2 on a sheet of paper, and the experiments are done accordingly. Let experiment $\mathcal{E}^1$ correspond to the number 1 on the paper and experiment $\mathcal{E}^2$ to the number 2. Select then one of the numbers randomly, and call the resulting experiment $\mathcal{E}$. Then information about the mean can obviously be obtained from this randomized experiment. However, in this case one can say that such information also can be found by selecting the numbers 1 or 2 in any nonrandom way because the choice of individual is, at least in a certain sense, random in this case also. If one agrees with this, the conditionality principle is valid in this case also, but for a different reason than in the original experiment: Here all 3 experiments give information about τ.

Assume next that this last variant of the experiments is done by an experimentalist A, but that there is another scientist B who possesses detailed information about the two individuals. Then one can argue that the inference done is valid for A, but not so for B. One can even argue that the validity of A's inference can be destroyed if B tells him what he knows. Similar discussion can again be done for the larger sampling case of selecting n out of N experiments. It also holds if the inference possessed by B is only partial, say that he knows that the selected persons have a higher avarage initial weight than the others. This may be a knowledge that he has obtained through a third scientist C who knows the complete identity of the persons behind the numbers and also their weights. In that case it might be that the inference may be adjusted in some way by taking the information given by B into account.

The problem bears some relation to problems one has in randomized field experiment when the randomization itself turns out to give a peculiar result. In such cases there are no simple answers to the question of when experimental results still can be considered to be valid. If the deviation in the randomization is relatively small, one can still rely upon the basic symmetry intrinsic in the randomization procedure.

In a similar way, even in cases where there is some knowledge present about the single experiments, as in Example 2.13.2, one can rely on the basic symmetry between experiments and choose to use unconditional inference. In this limited sense, we will not consider the conditionality principle to have absolute validity, at least in a practical sense, even though we hold it to be valid in simple situations.

In the light of this conclusion, we choose not to go into a detailed discussion of other principles like the likelihood principle, on which there is a large statistical literature; see [25, 28, 76] for a derivation of the likelihood principle from the conditionality principle. However we formulate it explicitly for completeness.

The likelihood principle. [25] *All information about the parameter θ obtainable from an experiment is contained in the likelihood function for θ given the observations y. Two likelihood functions for θ (from the same or different experiments) contain the same information about θ if and only if they are proportional to one another.*

By some, the likelihood principle is said to imply that Bayesian inference is the only valid one. In fact even a Bayesian inference where the prior depends on the experiment chosen, see [25] p. 21, is also doubtful according to this principle. It is the position of the present author that neither of these general principles, which may have rather strange consequences, can be said to have a complete general validity, at least from a practical point of view in an inference situation. Thus, in this connection I choose to be a little 'unprincipled'; see leCam [25]. It is an essential point that this whole discussion is carried out within the framework of the standard statistical experiment, a framework that one in my opinion sometimes have to transcend in applied statistics.

Finally I comment on the question of non-uniqueness of an ancillary, i.e., a statistic whose distribution is independent of the parameter. From a normative version of the conditionality principle, this leads to several possible ways to condition. There are many examples to illustrate this. Here are two simple ones.

Example 2.13.4. Let θ be a scalar parameter between -1 and +1. Consider a multinomial distribution on four cells with respective probabilities $(1 - \theta)/6$, $(1 + \theta)/6$, $(2 - \theta)/6$ and $(2 + \theta)/6$, adding to 1. Let the corresponding observed numbers in a sample of size n be y_1, y_2, y_3 and y_4. Then each of the statistics

$$a_1 = y_1 + y_2 \quad a_2 = y_1 + y_4 \tag{2.64}$$

is ancillary for θ, but they are not jointly ancillary. And conditioning upon a_1, respectively a_2 leads to distinct inference.

Example 2.13.5. In a certain city the sex-ratio is 1 : 1, and it is known that exactly 1/3 of the population have their own mobile telephone. The ratio between male and female mobile telephone owners is an unknown quantity $(1 + \theta)/(1 - \theta)$, where $-1 < \theta < 1$. One is interested in estimating θ by sampling randomly n persons from a register of the city population. It is assumed that the population is much larger than the sample size n.

Let the number of men in the sample be u_1, and let u_2 persons in the sample be owners of mobile telephones. Finally, let y be the number of male mobile telephone owners in the sample. The information (u_1, u_2, y) is enough to reconstruct the whole 2×2 table of frequencies. The entries in this table will have a multinomial distribution as in (2.1) with probabilities found from the information in the previous paragraph. Here both u_1 (being binomial $(n, 1/2)$) and u_2 (which is binomial $(n, 1/3)$) are ancillary, but they are not jointly so. The joint probability distribution of these two variables given by the frequency function

$$\sum_y C(y)(\frac{1+\theta}{6})^y(\frac{2-\theta}{6})^{u_1-y}(\frac{1-\theta}{6})^{u_2-y}(\frac{2+\theta}{6})^{n-u_1-u_2+y},$$

where $C(y)$ is the multinomial coefficient

$$C(y) = \frac{n!}{y!(u_1 - y)!(u_2 - y)!(n - u_1 - u_2 - y)!},$$

will namely always depend upon the unknown parameter θ.

The experiment can be conditioned in two different ways. First we condition upon the number u_1 of men. This variable can in principle have any integer value from 0 to n, but once it is determined, it is fixed, and the response variable y is binomial $(u_1, (1+\theta)/3)$, which gives an approximate inference for θ. More accurately, this fact together with the fact that the number $u_2 - y$ of female mobile telephone owners has an independent binomial $(n - u_1, (1 - \theta)/3)$ distribution implies that the likelihood is proportional to

$$(1 + \theta)^y (2 - \theta)^{u_1 - y} (1 - \theta)^{u_2 - y} (2 + \theta)^{n - u_1 - u_2 + y}, \tag{2.65}$$

which can be used to draw inference on θ, in particular to find the maximum likelihood estimator. Note that in this equation u_1 is fixed, while u_2 is a random variable. In particular, the Fisher information about the parameter θ is

$$\frac{u_1}{(1 + \theta)(2 - \theta)} + \frac{n - u_1}{(1 - \theta)(2 + \theta)}. \tag{2.66}$$

Alternatively, one can condition upon the number u_2 of mobile telephone owners. Then the response y, the number of men among these telephone owners, will be binomial $(u_2, (1+\theta)/2)$, and independently the number $u_1 - y$ of men among the non-owners of mobile telephone will be binomial $(n - u_2, (2 - \theta)/4)$. This will again give a likelihood proportional to (2.65), but with a completely different interpretation: Now u_2 is fixed, but u_1 is a random variable. The maximum likelihood estimate of θ will be the same as before, but the Fisher information for θ will be

$$\frac{u_2}{1 - \theta^2} + \frac{n - u_2}{4 - \theta^2}. \tag{2.67}$$

Finally, the unconditional experiment implies that $(y, u_1 - y, u_2 - y, n - u_1 - u_2 - y)$ has a joint multinomial distribution. Again the likelihood will be proportional to (2.65), but now both u_1 and u_2 are random. Again of course the maximum likelihood estimator will be the same, but the Fisher information will be different, namely

$$\frac{(2 - \theta^2)n}{(1 - \theta^2)(4 - \theta^2)}. \tag{2.68}$$

Asymptotically, using $u_1 \approx n/2$ and $u_2 \approx n/3$ the three expressions (2.66), (2.67) and (2.68) are approximately equal as one should expect, since this corresponds to the perfectly balanced case, but for finite u_1 and u_2 there is a striking difference.

There have been many attempts to find general rules for the choice of ancillary to condition upon in such cases, see for instance Cox [49]. My own view, in agreement with what I have expressed elsewhere in this book, is that we just have to live with the fact that the world can be seen from different perspectives, and that this sometimes may lead to different conclusions.

2.14 A Few Design of Experiment Issues

As has been said before, a statistical investigation begins with a question, or often, one and more questions. Unfortunately, many statistical textbooks and introductory courses give the impression that inference is just tied to a class of probability measures, and the whole issue of experimental design is relegated to later courses, if at all. Fortunately, there exist good, rather simple, books on experimental design like Montgomery [154] and Box, Hunter and Hunter [34]. A treatise of very many aspects of experimental design, including both practical issues and mathematical issues, will be given in the forthcoming book by Bailey [15]. This is not the place for a systematic discussion of the topic of design, but I want to stress as a general point that any empirical investigation starts with a design phase, and that any full discussion of empirical data should take into account the phase where questions were posed in the first place.

The question that is posed in an experiment can typically be related to treatments that can be applied to a set of units. The units can be human beings, animals, plants, plots in a field or runs in an industrial process.

Historically, the first modern discussion of experimental design was given by R.A. Fisher; see [84]. Fisher stressed 3 basic principles.

1) Replication
This means that every treatment should be repeated on several units. First, this is a prerequisite for inference in many cases. Secondly, it is a precautionary measure if something should go wrong.

2) Randomization
This is the cornerstone underlying the use of statistical methods in practice. Randomization means that both the allocation of the experimental material and the order in which the individual runs or trials of the experiment are to be performed, are randomly determined. This may serve to validate the statistical model, and is important for the validity of statistical tests.

3) Blocking
Blocking is a technique used to increase the precision of an experiment. A block is a portion of the experimental material that should be more homogeneous than the entire set of material. In an ordinary block experiment, every treatment is applied once to every unit in each block.

I will give a brief sketch of the coupling between the experimental design phase and the statistical analysis phase from our point of view. Many details are left out.

Consider a set Z of potential experimental units for some experiment; this set can be finite or infinite, one may even consider an uncountable number of units. For each given $z \in Z$, let y_z be some potential response variable, and let μ_z be the expectation of y_z for the case where no treatment is introduced.

One may also have a set T of potential treatments which can be applied to each unit. Let μ_{tz} be the expectation of y_z, given z, when treatment t is applied to z, and define $\theta_{tz} = \mu_{tz} - \mu_z$. Assume for simplicity that the y_z's are independent with variance σ^2. Let η_z denote other parameters connected to the unit z.

In this situation it is natural to call $\phi = (\{\mu_z, \eta_z; z \in Z\}, \{\theta_{tz}; t \in T, z \in Z\}, \sigma^2)$ a c-variable for the system and $\Phi = \{\phi\}$ the c-variable space. This terminology is consistent with what I have used elsewhere, and it is also consistent with the terminology I will use in my approach to quantum mechanics later. Note that ϕ is not estimable in any conceivable experiment; nevertheless, it is a useful conceptual quantity.

Now in the experiment itself one selects some finite subset Z_0 from Z. I will assume that this is done in such a way that $\theta_t = E(\theta_{tz}|t, z)$, expectation over this selection mechanism, is independent of the selected z. Then we will have for the given selected unit $z \in Z_0$ that

$$E(y_z|t, z) = \mu_z + \theta_t. \tag{2.69}$$

This is one way to express the well known unit/treatment additivity, which is considered by Bailey [13, 14] and others as a prerequisite for having a consistent approach to the design of experiment. The parameter of interest then will typically be $\theta = \{\theta_t; t \in T\}$, which in principle is some function of ϕ, in practice a very complicated function, but this does not matter.

I will come back to this description of experiments, and in particular to the randomization of such an experiment in Chapter 3 and Chapter 7.

2.15   Model Reduction

Consider now a standard statistical experiment for which we first contemplate a full parameter θ and perhaps a model Q^θ relative to this parameter. As a matter of principle, a parameter is defined through some linguistic expression, and sometimes it may be desirable or even necessary to use a simpler language and hence a simplified parameter $\lambda = \lambda(\theta)$. In certain cases the original parameter θ may not be estimable in any possible experiment, and then the only way to get some information at all, will be to limit oneself to such a parameter λ. In this case, the part of θ not contained in λ can be said not to have a value at all. We thus take the position that if nothing can be said about a parameter from any experiment or in any other way, we might as well say that this parameter does not have a value.

We will later meet an extreme example of this kind of behaviour in the electron spin system. For such cases the original contemplated model Q^θ does not make sense; it is only possible to speak about a reduced statistical model Q^λ.

A less extreme case is when it pays in order to improve prediction performance to reduce the statistical model in such a way.

Example 2.15.1. Consider the elementary linear regression model $y_i = \beta_0 +$

$\beta x_i + e_i$, where the error terms e_i are independent and identically normal $(0, \sigma^2)$, and where we for simplicity assume that the values $x_1, \ldots, x_n$ of the explanatory variable satisfy $\sum x_i = 0$. Then the mean squared prediction error at the fixed point x_0 under the nonrandom explanatory variable model can be shown (see Appendix A.1.6) to be

$$PE = \frac{\sigma^2}{n} + x_0^2 \frac{\sigma^2}{\sum x_i^2} + \sigma^2. \tag{2.70}$$

The present model has a small number of parameters, but even here model reduction can be considered. The strength of the example is that an explicit solution is easy to find. A natural reduced model here will be one with slope 0, leading to the prediction error implied by using the reduced model when data are generated by the full model (Appendix A.1.6 again):

$$PE^R = \frac{\sigma^2}{n} + \beta^2 x_0^2 + \sigma^2. \tag{2.71}$$

Comparison with (2.70) shows that in terms of prediction error, model reduction pays (for all non-zero x_0) if and only if

$$\beta^2 < \frac{\sigma^2}{\sum x_i^2}. \tag{2.72}$$

This condition can also be written as $|t| < 1$, where t is the 'theoretical t-value' $\beta/\text{std}(\hat{\beta})$ with $\text{std}(\hat{\beta}) = \sigma/\sqrt{\sum x_i^2}$. In this form it can be shown that the condition also can be generalized to the question of deleting a single variable from a multiple regression model, a fact that also has been referred to in applied statistics books like Snedecor and Cochran [189].

In Chapter 3 we shall come back to this and then mainly concentrate on the random x regression, where the assumption on the error terms is that ϵ_i, given all the x-variables are independent and identically normal $(0, \sigma^2)$, and we typically may assume that $x_1, \ldots, x_n$ are independent and identically normal $(0, \sigma_x^2)$. Thus the assumption $\sum x_i = 0$ is replaced by $\text{E}(x_i) = 0$. Then the new prediction error is found by taking the expectation over the x-variables in (2.70) and (2.71), leading to the following criterion for model reduction:

$$\beta^2 (n-2) < \frac{\sigma^2}{\sigma_x^2}. \tag{2.73}$$

This illustrates explicitly the statement made earlier that model reduction may be beneficial when the number of data points is small or moderate. Unfortunately, the criterion depends upon unknown parameters - a general problem in this area.

A very general large sample discussion of model reduction using likelihood theory can be found in Hjort [122, 123]. A number of specific examples of model reduction can also be found in that paper and in references given there. In Chapter 8 we will go through a large example of model reduction of the type illustrated by Example

2.15.1. A prerequisite for doing model reduction of this kind is that there is an appreciable model error σ^2 in the situation, or that the estimation in the original model is unstable. In the case of the linear model $y = X\beta + e$ this is equivalent to near collinearity $(\det(X'X) \approx 0)$. In fact, these two criteria is consistent with what one can read from the simple inequality (2.72).

Thus we have sketched two situations where model reduction is called for: 1) Parts of the parameter do not have a value relative to the experiment; 2) We want to improve prediction error. In fact, these two situations are related. We can see this in the case of the linear model, where the first situation corresponds to exact collinearity $(\det(X'X) = 0)$ and and the second one to near collinearity.

2.16 Perfect Experiments

From a physicist's point of view, the bulk of both this chapter and the next chapter is about measurement apparata. Both in physics and in other sciences, the following is true: For a given state of nature one can choose to use each of many different potential apparata to observe or measure the state. The state variable θ must be the same irrespective of which measurement apparatus one uses, but the measurement model Q^θ will of course depend upon the apparatus.

In some instances in physics, and also in other sciences, one can disregard measurement error. In this situation one has a perfect measurement, and the measurement model will then give Q^θ as a delta-measure. For a continuous state variable θ this must be regarded as an approximation, but, as discussed at the end of Section 2.4, if θ or its associated reduced parameter λ takes a discrete set of values $\lambda_1, \lambda_2, \ldots$, then it is realistic to assume that a measurement apparatus can be almost completely perfect.

But even in the case of a perfect measurement something can be learned from the discussion in this chapter. A state is not in general just characterized by a simple number. The parameter θ (or λ) is the answer of a given, specific question to nature. And as such:

1) θ may be part of a c-variable ϕ to which symmetry arguments apply.

2) θ may depend upon the context for the measurement/experiment.

3) Even though we have stated (Section 2.13) that the conditionality principle in our view should not always have complete general validity, the principle could be relevant and valid for the situation at hand, and then statements about the parameter should be limited to the experiment actually performed.

4) A suitable model reduction may be called for, even though this at first does not seem to be particular relevant for a situation where we are able to get perfect information about the parameter. However, for presentation of a complicated result it may still be desirable.

Chapter 3

STATISTICAL INFERENCE UNDER SYMMETRY

3.1 Introduction

[1]The main issue in this chapter is to explore the consequences of adjoining a symmetry group to a statistical model. This will be the platform for much of the discussion in later chapters. At the outset, the group will be introduced as a transformation group on the parameter space or on the space of observations. A transformation group on the parameter space will be particularly important. Statistical inference, as we have introduced it in the previous two chapters, has to do with deducing statements on a parameter or on a set of parameters. So, if we arrive at a certain conclusion concerning one parameter, symmetry may also lead to a similar conclusion about a transformed parameter.

This whole issue may also be related to the foundation of probability: One can roughly imagine three possible such foundations: One based upon limits of empirical frequencies as the number of observations tends to infinity, one based upon subjective judgement, and one based upon symmetries like in die tossing or in card games. Then one can imagine in a similar way three corresponding possible bases for statistical inference: The large sample basis, the Bayesian basis and the group theory basis. The latter is not always applicable, but when it is, it may be quite powerful. Later, in Section 3.6, I will discuss a tentative relationship between the three approaches.

One topic of importance will be the idea of model reduction under symmetry. I will argue that every sensible model reduction should be to an orbit or to a set of orbits of the group acting upon the parameter space. This will turn out to have important consequences, both when we in later chapters discuss quantum mechanical modelling and when we discuss models of use in chemometrics.

Group actions in the statistical model setting are introduced in Section 3.2.

---

[1]This chapter is fairly mathematical. Those who feel uncomfortable with this mathematics, can in the first reading look comparatively quickly through the results, although some of these are essential to understand in order to appreciate the main development and the connections to parts of the rest of the book. As a very minimum, the introduction together with the summary of each section should be read. Needless to say, those wanting to understand the theory, should read all details.

The necessary concepts are then developed in the Sections 3.3 and 3.4. Much of the remaining chapter is related to estimation, but there is also much related discussion, as on model reduction in the Sections 3.7 and 3.8, and a demonstration in Section 3.6 that the introduction of an appropriate group structure may diminish the distance between Bayesian inference and classical inference. Part of Section 3.8 is a preparation for Chapter 8, and Section 3.9 introduces an important estimation method in mixed model analysis, while at the same time illustrates the technique of estimating the orbit index in the parameter space from the orbit index in the space of observations.

As recently pointed out by Breiman [36], the theoretical statistical literature is to a large extent dominated by aspects of probability-based data models. That paper initiated an interesting discussion, but whatever attitude one should have here, it is highly relevant to ask whether also other structural elements can be of importance for statistical methodology, elements which in terms of their practical implications perhaps so far have not been sufficiently focused upon in the literature. In this chapter we will mainly study symmetry aspects by letting a group G act upon the parameter space Θ and the sample space S. This is of course not a new concept, but the tendency in the statistical literature has been to start with a model, and let the group be induced by this model. Here we want to argue for the group as an independent entity in addition to the model. The introduction of such a group will be shown to have several consequences for the analysis of data.

Throughout most of this chapter, I will use a standard statistical model as introduced in Chapter 2, but I will augment this model by introducing a group G acting upon the sample space and correspondingly on the parameter space. In Section 3.7, I will go beyond the standard model when looking at model reduction, and in Section 3.11, I go beyond the standard model and look at an extended parameter space, a c-variable space, which may include parameters of several potential experiments.

The first issue then is the choice of the group G. We often talk about a symmetry group, but in fact rather weak symmetry requirements are needed: I will demand that the sample space S is closed under the actions of the group and that the parameter space Θ is closed, too. These actions are written:

$$y \mapsto yg; g \in G, \text{ respectively } \theta \mapsto \theta g; g \in G.$$

(The use of the same symbol for the transformations on both spaces will be discussed later. For reasons that will be clear in Section 3.3, I will throughout the book place the group symbol to the right of the object it acts upon.)

Then we essentially only need the property that for any group element $g \in G$ we should have that $\{yg, \theta g\}$ may be considered to be an equally natural basis for inference as the original data and parameters $\{y, \theta\}$. Roughly, also, the symmetries expressed by the group should have some substantial basis in the concrete problem described. The introduction of a group in this way has for instance earlier been proposed by Fraser [86, 87].

These requirements may lead to several possible choices for G in concrete cases.

Berger [24], among others, consider this as an disadvantage, and proposes to consistently use the least group which is transitive on the parameter space. (The concept of a least group may be made precise by considering subgroups.) We will later encounter sensible groups that are not transitive on the parameter space, however. My own general position is that we sometimes just must live with the fact that we have a choice between several useful groups, as we may have the choice between several good models or between sensible loss functions.

A very simple transformation group may be the change of units for the observations y: $yg = by$, for instance changing from meter to centimeter. Another simple transformation group may be described as changes of origin: $yg = a + y$. Other transformation groups will be introduced shortly.

Turn first to the Bayesian approach. An important problem here is to specify a prior distribution for the case where a priori we have no information about the parameter of the statistical model. The literature varies somewhat with respect to what should be meant by a non-informative prior (see the comprehensive review by Kass and Wasserman [134]). When this term is used in situations with symmetry (in the simplest case location and/or scale symmetry), I claim that the non-informative prior may be taken to be the right invariant measure on the transformation group, a measure that can be defined in a natural way for most well-behaved groups. Definitions and arguments will be given in Section 3.3.

I will argue from several points of view that when one has prior symmetry information given by some fixed group, then this information should be made explicit by also using other quantities connected to the group, not only the invariant measure. This proposal is in conflict with the Bayesian view that all prior information should be expressible as a measure.

Finally, I will discuss the concept of model reduction, and relate it to the orbits of the group as acting upon the parameter space. This requires that we first discuss the concept of orbits.

3.1.1 *Orbits*

Recall from Definition 1.10.2 that an orbit in the parameter space Θ is defined as any set of the form $\{\theta_0 g\}$, where θ_0 is some fixed point in Θ. The whole parameter space is always divided into a disjoint set of orbits. If the space consists of only one single orbit, we say that the group action is transitive. In a similar way we define orbits of the group acting on the sample space.

The concept of orbit will be so central in the discussions of this book that we already now give some simple examples:

Example 3.1.1. Let Θ be a finite dimensional vector space, and let G be the group of rotations in this space: $\theta g = P_g \theta$ for orthogonal matrices P_g. Then the orbits of G are the spheres with center 0, i.e., the sets of vectors with fixed norm.

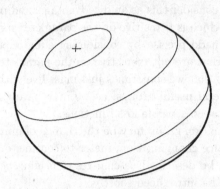

Fig. 3.1   Orbits on the earth.

The one point set $\{0\}$ is one particular orbit.

Example 3.1.2. Let the parameter space Θ be a sphere like the surface of the earth, and let the group elements be given by rotations around some fixed axis. Then the orbits will be latitude circles relative to this axis.

Example 3.1.3. Let Θ be a p-dimensional vector space, and let the group G be defined by $\theta g = \theta + a_g$ for arbitrary p-vectors a_g. Then the group will be transitive, that is, there is one orbit, the whole space. A much more restricted case is the following: Let $\theta g = \theta + b_g a$ for arbitrary scalars b_g and a fixed p-vector a. In this case the orbits are all lines parallel to a.

Example 3.1.4. Let $\mathcal{X}$ be the space of $n \times p$ matrices X, where $n > p$. Let G be defined by rotations in both the column space and the row space here: $Xg = P_g X Q'_g$, where P_g and Q_g are orthogonal matrices and $'$ denotes transpose. The orbits of this group are most easily found by using the singular value decomposition: Each X can be uniquely written as $X = P_0 \Lambda^{\frac{1}{2}} Q'_0$, where $\Lambda = (\Lambda'_0, 0')'$, and Λ_0 is a diagonal matrix with the ordered eigenvalues of $X'X$ on the diagonal. Again P_0 and Q_0 are orthogonal. This implies that $Xg = P_g P_0 \Lambda^{\frac{1}{2}} (Q_g Q_0)'$, and hence the orbits of G in $\mathcal{X}$ are characterized by the ordered set of eigenvalues of $X'X$.

Example 3.1.5. Let Θ be a large parameter space for some scalar observation, with the only prerequisite that the parameter space contains a location parameter μ and a scale parameter σ in such a way that it is closed under the scale and location transformations given by $(\mu, \sigma)g = (a_g + b_g \mu, b_g \sigma)$, and that the class of distributions indexed by θ shall contain the normal distributions. This then corresponds to some large family of distributions of our observations.

I will argue more in detail in Section 3.7 that any model reduction should be to an orbit or to a set of orbits of the underlying group acting upon the parameter space. One obvious general argument should be clear already now: Also after model reduction, it is natural to demand that the parameter space is invariant under the group. Of course not all such reductions are sensible, they must be checked by independent, complementary means, and in particular they must be checked against observations, but these are the only candidates that we will consider.

Here is then one obvious candidate for model reduction in the case of Example 3.1.5: Look at the subclass of Θ giving the normal distributions with some unspecified expectation μ and some unspcified standard deviation σ. This subclass is closed under the scale and location transformation, and is easily seen to constitute an orbit for this group in Θ. This can be considered to be an argument for the normal approximation. Of course there are other location and scale families of distributions, and these will also constitute orbits under the group given here. But then there are other arguments for the normal approximation in the literature, and taken together they seem quite convincing. The particular argument given here, seems to be new.

Example 3.1.6. The same argument applies to just the location group given by $(\mu, \sigma)g = (\mu + a_g, \sigma)$, where the set of normal distributions constitute a set of orbits, one for each σ, and for the scale group $(\mu, \sigma)g = (b_g\mu, b_g\sigma)$, where the set of normal distributions again constitute a set of orbits, one for each value of μ/σ. The same type of argument can also be generalised to the multivariate case, then as an argument for a model reduction to a multivariate normal distribution with an expectation vector μ and a covariance matrix Σ. In fact, this is an important case; very much multivariate methodology has been developed for the multinormal case, and the fact that this methodology can be justified strictly after a proper model reduction, serves to give it an extended validity. A similar argument will be used in Chapter 8.

In Examples 3.1.1, 3.1.2 and 3.1.3 and in all similar examples of groups acting on some parameter space the orbits are characterized by a certain parameter in the space, the *orbit index* τ. Technically, we must assume the group to be *proper*, a term which is defined in Appendix A.2.1. Then the orbits are closed sets in the topology of Θ, and the orbit index can be defined as follows.

Definition 3.1.1. *An orbit index is any subparameter τ such that the orbits are given by a fixed value of τ.*

In Example 3.1.1 the orbit index is given by the radius of the sphere. In Example 3.1.2 the orbit index is simply the latitude. In the last part of Example 3.1.3 the orbit index may be taken to be an arbitrary vector perpendicular to the fixed vector a.

The orbit index can alternatively be characterized as a *maximal invariant*: τ is invariant under the actions of the group, and the maximal parameter with this property.

For simplicity we repeat Definition 1.11.3 in this language.

Definition 3.1.2. *A transitive group has only one orbit, and therefore only a single, constant value for the orbit index.*

In general there is an optimal (best equivariant; Bayes using right invariant prior) estimator on each orbit. This is our main argument for always confining model reduction to be to one or several orbits of the group. For details, see Sections 3.5, 3.7 and 3.8 below.

One can also consider orbits in the sample space. The orbit index here is often given by a variable a; see the next Section.

3.2   Group Actions and Statistical Models

Summary: In this section I introduce group actions on the parameter space and on the sample space of a statistical model. This is illustrated by several examples. The concept of stabilizer is defined. The orbit index in the sample space is introduced, and it is proved that the distribution of the orbit index in the sample space only depends upon the orbit index in the parameter space.

I will take as a point of departure the standard statistical situation. Hence, let a statistical model be defined as a class of measures $Q^\theta(dy)$ on a sample space S, where the parameter θ lies in the parameter space Θ. Now introduce a symmetry group G into this setting.

Let us start with the sample space S. A transformation group G is assumed to act on this space by $y \mapsto yg; g \in G$. Again recall the convention of placing the group symbol g after the element to be transformed. Like other authors in this field, e.g., Dawid, Stone and Zidek [54], I will not focus much on topological and measurability questions, but some issues and references are mentioned in Appendix A.2.1.

With a model as above, this also automatically induces a class of transformations g on the parameter space. We can namely define θg by

Definition 3.2.1.

$$Q^{\theta g}(A) = Q^\theta(Ag^{-1}). \tag{3.1}$$

In order that (3.1) shall determine uniquely the transformations $\theta \mapsto \theta g$, we must require that the parameter of the model should be identifiable; see Definition 1.9.1. I will always assume this. One must also assume that the class of measures defined by the righthand side of (3.1) defines elements of the original parametric model, in short that the parameter space Θ is closed under actions g.

Since we in Section 2.2 defined the model in terms of a class of underlying probability measures as $Q^\theta(A) = P^\theta(y \in A)$, the definition (3.1) implies that

$$Q^{\theta g}(A) = Q^\theta(Ag^{-1}) = P^\theta(y \in Ag^{-1}) = P^\theta(yg \in A).$$

I will explore consequences for statistical inference later, but already from this we can say that estimators ought to have some invariance properties.

Equation (3.1) also implies what mathematicians call a *homomorphism* between the two group actions: If g_1 and g_2 act on the sample space and one then introduces similar actions on the parameter space by (3.1), the product $g_1 g_2$ is mapped by (3.1) in a consistent way. Or to be more precise, if we start by using a different notation for the two group actions, these two group actions behave in exactly the same way, so one might as well use the same notation. That is:

Lemma 3.2.1. *Given a transformation group G acting upon the sample space S, introduce the actions $\bar{g}$ on the parameter space Θ by for each $g \in G$ and each θ defining a measure $Q^{\theta\bar{g}}$ by $Q^{\theta\bar{g}}(A) = Q^\theta(Ag^{-1})$. Then $\overline{g_1 g_2} = \bar{g}_1 \bar{g}_2$. This forms the homomorphism property. As a consequence, the unity e transforms as it should, and $\overline{g^{-1}} = \bar{g}^{-1}$.*

Proof:

$$Q^{\theta\bar{g}_1\bar{g}_2}(A) = Q^{\theta\bar{g}_1}(Ag_2^{-1}) = Q^\theta(Ag_2^{-1}g_1^{-1}) = Q^\theta(A(g_1 g_2)^{-1}).$$

The rest follows since the elements $\{\bar{g}\}$ with the product defined form a transformation group on Θ.

On the background of these results, I choose in this book to use the same symbols g, G for transformations on both spaces. Some authors like Berger [24] choose to use different symbols, but my terminology is simpler, and it is consistent with the one pure mathematicians use when discussing group actions.

A pure mathematician will think of G as an abstract set of elements with a multiplication table. As statisticians we are more interested in G as a transformation group: It introduces an action $\theta \mapsto \theta g$ on the parameter space, and similarly on the sample space by $y \mapsto yg$. For these transformation groups the usual group properties are fairly obvious:

Lemma 3.2.2. *For a transformation group the following is automatic: An associative multiplication $g_1 g_2$ is defined between all group elements by considering two successive transformations, and there is a unit $e \in G$ and an inverse g^{-1}.*

In principle the statistical model implies comparatively few restrictions on the choice of group: As discussed in Section 3.1 the essential mathematical requirement is only that the full spaces S and Θ should remain invariant under the transformations, and that $(yg, \theta g)$ is a starting point as natural as (y, θ). Later we will require that the loss function shall be invariant under the group and that the right invariant measure of the group on Θ should be a natural non-informative prior to use.

Nevertheless the choice of a transformation group may often be related to the selected statistical model. Several authors, incuding Dawid [52] have used group theory and group representation theory to characterize factorial models and their associated hypothesis structure. Some other examples follow.

Example 3.2.1. Here is an example where the same model class in a very natural way can be endowed with very different transformation groups according to the situation: Many statistics textbooks make a point of the fact that both regression and analysis of variance problems can be handled by similar linear models of the form $y = X\beta + e$. Nevertheless, most applied researchers feel strongly that there are essential differences between the two situations. From our point of view the difference is clear: In the regression situation it is natural to use the linear group $\beta \mapsto A\beta + a$ or some subgroup. In the case where all y-variables are measured in different units, it may be relevant to go to the subgroup defined by $\beta_j \mapsto k_j\beta_j; j = 1, ..., p$ for the regression components, and when the units are orthogonal, we will later make use of the subgroup obtained by restricting the matrix A to be orthogonal. On the other hand, in the analysis of variance situation the natural group is some permutation group, the choice of group depending upon the randomisation used. (We will say more about this in Section 3.10 below.)

Example 3.2.2. Simple groups in any data situation with some scalar data y are the translation group defined by $yg = y + a_g$, the scale group $yg = b_g y$, where $b_g \neq 0$, and the translation- and scale-group given by $yg = b_g y + a_g$. One may also imagine extensions where a_g and/or b_g are given different values for data from different sources. Another common group is the rotation group in multivariate analysis.

Example 3.2.3. Note that the concrete actual appearance of the group actions on the sample space and on the parameter space may be different. Here is an example: Let the multivariate data (X, y) have rows which are independent and identically distributed with x-covariance matrix Σ_{xx}, conditional y-variance $\sigma^2 = Var(y_i|x_i)$ and (x, y)-regression vector β. Let G be defined by rotations in the x-space: $(X, y) \mapsto (XC_g, y)$ for orthogonal matrices C_g. Then the induced transformations in the parameter space are $(\Sigma_{xx}, \beta, \sigma) \mapsto (C_g'\Sigma_{xx}C_g, C_g'\beta, \sigma)$, apparently something different from the sample space group action.

Above I started with actions on the sample space. In other cases it is more natural to define the group actions directly on the parameter space. Even then any model must have the property that there also can be defined group actions on the sample space, and this in such a way that (3.1) holds. In general this does not lead to a unique action upon the sample space, but it is shown in Proposition 3.11.2 below that if we regard the sample space as the space of a complete sufficient statistic, then the sample space group action is indeed uniquely determined from the group action on the parameter space.

Look again at the group of transformations acting on the sample space. For a given point $y_0 \in S$, let the orbit generated by this point be $\{y : y = y_0 g \text{ for some } g\}$. These orbits constitute equivalence classes in S; so we can (under weak assumptions; see Appendix A.2.1) index the classes by some random variable a: Different values of a corresponds to different orbits of the group G in the sample space S. Alternatively, the random variable a can be regarded as a *maximal invariant* in the sample space: It is invariant under the actions of G, and any other invariant stochastic variable can be written as a function of a.

Definition 3.2.2. *If for every pair of points y_1 and y_2 there is not more than one group element g which transforms y_1 into y_2, we say that the group is* free. *This means that the group transforming a given element into itself is trivial; in general this group is called the* stabilizer. *Stabilizers S_y for different elements y may be transformed in a simple way into each other in the transitive case: $S_{y_2} = g^{-1} S_{y_1} g$ if $y_2 = y_1 g$.*

If the group is both transitive and free, then one can pick one arbitrary basis point y_0 and write every element y in a unique way as $y = y_0 g$. Thus in this case there is a one-to-one correspondence between the group and the sample space.

We have already defined orbits and the maximal invariant for the group acting upon the parameter space, and we may also, in a similar way, introduce the concepts of free group and stabilizer for the actions on that space. Hence on the parameter space also we can have either of four basically different situations: free or not free/transitive or not transitive transformation group. Since a statistical model usually implies a simplification of the data structure, one might say roughly that the parameter space Θ typically is 'smaller' than the sample space S. Hence, it is more common for the action of a group to be transitive on the parameter space than on the sample space.

As already mentioned, under very weak assumptions, formulated in the Appendix, the orbits of a group can be given an index a in such a way that different orbits have different values of that index and conversely. Using (3.1) it is straightforward to prove the following:

Lemma 3.2.3. *The distribution of the orbit index a in the sample space depends only upon the parameter τ, the orbit index in the parameter space. In particular, if the parameter group is transitive, then the orbit index a is ancillary, i.e., has a distribution which is independent of any model parameter.*

Proof.
It follows from the definition that $a(yg) = a(y)$, since a is invariant under the group G. Therefore

$$P^{\theta g}(a(y) \in A) = P^{\theta}(a(yg) \in A) = P^{\theta}(a(y) \in A).$$

Thus this probability is invariant under the transformation $\theta \mapsto \theta g$, and must be constant on orbits in the parameter space.

3.3   Invariant Measures on the Parameter Space

Summary: In this section I first introduce the right Haar measure and the left Haar measure on the group itself. From this, the concept of relative invariant measure on the parameter space is defined, with special cases the right invariant measure and the left invariant measure. A list of arguments is given in order to argue that the right invariant measure quite generally should be chosen as the solution when an informative prior is needed in Bayesian analysis. The construction of the right invariant prior is described in more detail.

An important issue is that the choice of a symmetry group in a statistical setting also implies a natural choice of a non-informative prior distribution. In certain cases it is obvious what this invariant prior will be. For example, for the translation group given by $\theta \mapsto \theta + a$ for a scalar parameter θ we get in a natural way $\nu(d\theta) = d\theta$, while the rotation group for a p-dimensional vector θ has the natural rotation measure: Uniform distribution on the p-dimensional unit sphere.

First we look at measures on the group itself. It is well known (Nachbin [158]) that every locally compact group G possesses a socalled *right invariant Haar measure* ν_G on this group, that is, a measure with the property that $\nu_G(Dg) = \nu_G(D)$ when $g \in G$ and D is a set in the space of group elements, where we define $Dg = \{hg : h \in D\}$.

The right Haar masure is unique except for a multiplicative constant; see [158] or [121], ch. IV. When G is compact, ν_G can be taken to be a probability measure.

There is also a dual *left invariant Haar measure* μ_G, i.e., a measure satisfying $\mu_G(gD) = \mu_G(D)$, where $gD = \{gh : h \in D\}$. This left invariant measure is equal to ν_G (except possibly for a multiplicative constant) for compact groups, if the group is finite or countable or if the group is commutative. In general the connection between the two measures is $\mu_G(dg) \propto \Delta_G(g)\nu_G(dg)$ for the socalled *modulus* or modular function Δ_G, whose properties are discussed in Appendix A.2.2.

Now turn to measures on the parameter space. If G is a group of transformations acting on this space, then our first recommendation concerning a non-informative prior ν on Θ is that it should be relatively invariant, that is, satisfy

$$\nu(\Gamma g) = \delta(g)\nu(\Gamma) \tag{3.2}$$

for some multiplier (multiplicative function) δ, where Γ is an arbitrary set in the parameter space, and $\Gamma g = \{\theta g : \theta \in \Gamma\}$.

Definition 3.3.1. *If the multiplier δ is identically equal to 1, so that $\nu(\Gamma g) = \nu(\Gamma)$, we say that the relatively invariant measure is a right invariant prior on Θ.*

The relatively invariant measure is only uniquely defined on orbits in the parameter space, so it is only unique on the whole space when the group action on Θ is transitive.

Existence of a relatively invariant measure ν can be shown to follow quite generally. In fact the existence of a right invariant measure on the parameter space follows under weak assumptions from the existence of the right Haar measure on the group itself. In certain cases one may also define a left invariant measure on the parameter space; see Appendix A.2.2.

For the following list of reasons, we will in this book recommend the right invariant measure as a prior in all cases where a group action is defined on the parameter space and a non-informative prior is needed. Note that this non-informative prior is only uniquely defined when the group action on the parameter space is transitive. Otherwise it must be accompanied by some measure on the orbit index τ.

(1) It is reasonable that the posterior measure at least should stay proportional if corresponding transformations are made of the sample space and the parameter space; we may let the constant of proportionality depend upon the group element. Simple arguments show that this - together with the invariance requirement on the model - implies that the prior must be relatively invariant.

(2) A relatively invariant measure on a group G acting on Θ may induce a relatively invariant measure on Θ. This can always be done in the right invariant case, otherwise it will only be possible if the stability group is trivial (or if the modular function is identically 1) (cf. Theorem A1).

(3) Recent results by Eaton and Sudderth [69–73] show that under very general conditions all other invariant inferences than those based upon right invariant priors are strongly inconsistent in a well-defined sense.

(4) Under invariant quadratic loss, the optimal equivariant estimator is the Bayes estimator with a right invariant prior (Corollary 3.5.5).

(5) In the transitive case, confidence intervals and credibility intervals corresponding to right invariant prior are numerically equal (Section 3.6).

(6) It is shown in [72] that the use of the right invariant measure yields the same invariant predictive distribution as certain other methods under specific assumptions.

(7) With a fixed group, and when inference is restricted to permissible parameters (see the next Section), the marginalization paradoxes of Dawid et al. [54] are avoided (see Helland [112]) when the right invariant prior is used.

(8) Posterior distributions of invariant joint functions of parameters and data will under certain conditions have the 'correct' sampling distribution (see references in [51]). For example, for independent identically normal (μ, σ^2) data under the translation- and scale- group, when the right invariant prior (density $\propto$

σ^{-1}) is used, the posterior distribution of the t-statistics will be a Student's distribution with $n-1$ degrees of freedom, but not so under the left invariant prior.

(9) When normalized priors converge to a right invariant measure, then the posteriors also converge as they should under weak assumptions [195].

(10) This choice of prior leads to a close link to Fraser's structural inference [54].

(11) There are links to other non-informative priors (Kass and Wasserman [134]),

As a balance against all these arguments, it should also be mentioned that there exist cases, admittedly rather extreme (non-amenable groups; see [32]) where a right invariant prior may lead to a uniformly inadmissible estimator [70].

I will consider this as overweighted by the long list of arguments for the right invariant prior.

The Bayes estimators that are obtained using a right invariant measure as prior will also have several other good properties, and could equally well have been arrived at by using these properties. When the loss function is quadratic and invariant under the group and the group action is transitive on the parameter space (see Subsection 3.1.1), the estimators will be best equivariant (see Section 3.4) under the group in question, and they will typically be minimax, that is, they minimize the maximum over θ of reasonable risk functions $R(\theta)$. (Recall from equation (2.36) that the risk function is the expectation of loss function under the model.)

The connection between ν_G on G and ν on Θ is relatively simple: If for some fixed parameter value θ_0 the function β is defined by $\beta(g) = \theta_0 g$, then $\nu(E) = \nu_G(\beta^{-1}(E))$, where ν_G is the right Haar measure on the group. This connection between ν and ν_G can also be written

$$\nu_G(dg) = \nu(d(\theta_0 g)). \tag{3.3}$$

The measure ν_G can often be calculated using a suitable Jacobi-determinant. When Θ is non-compact, which usually is the case, ν will be an improper prior, that is, the integral over ν is infinite.

I have already sketched the right invariant measure for some simple groups. Some other relatively simple cases are: For the scale group on a scale parameter $\sigma \mapsto b_g \sigma$ the invariant measure is $\nu(d\sigma) = d\sigma/\sigma$, while for the combined translation- and scale-group $(\mu, \sigma) \mapsto (b_g \mu + a_g, b_g \sigma)$ it is $\nu(d\mu, d\sigma) = d\mu d\sigma/\sigma$ (see, for instance Berger [24]).

In other cases the construction of ν is not so simple. The following general rule [24] is useful: Assume that the transformation group G on the p-dimensional parameter space Θ can be considered as a subset of $\mathbb{R}^p$ with positive Lebesgue measure. If $J_g(h)$ is the Jacobi-determinant for the transformation of G given by $h \mapsto hg$, and if e is the unit element of the group G, then $\nu_G(dg)$ will be a measure with density

$$f(g) = \frac{1}{J_g(e)}.$$

3.4   Subparameters, Inference and Orbits

Summary: In this section the concept of a permissible subparameter is defined: A subparameter such that the group actions can be transfered to it from the action upon the original parameter. First it is shown, among other things by considering a Bayesian reduction principle, that this is a fruitful concept. Next, several special cases are described. It is proved that every subparameter may be made permissible by going to a subgroup. Also the concept of an equivariant estimator is defined, and several uses of group symmetry in statistical inference are sketched.

Quite often in a statistical analysis, a subparameter, that is, some function of θ, is needed.

Definition 3.4.1. *Inference should preferably be limited to parametric functions that are permissible sub-parameters under the group G, that is, the parametric function $\eta(\cdot)$ should be such that $\eta(\theta_1) = \eta(\theta_2)$ implies $\eta(\theta_1 g) = \eta(\theta_2 g)$ for all $g \in G$.*

This requirement then allows G to act as a transformation group on the range of $\eta(\cdot)$, since there is a unique way of defining ηg by

$$(\eta g)(\theta) = \eta(\theta g). \tag{3.4}$$

This is an essential property: Group transformations are transferred in a natural way from the original parameter to a subparameter. The relationship between McCullagh's general category theory based requirement on statistical models and that of permissibility of subparameters is discussed on p.1241 of [156].

But the assumption of permissibility also turns out to have other desirable implications. First, this assumption implies under transitivity that credibility sets and confidence sets are numerically equal with the same associated probability/ confidence level; see Theorem 3.6.1 below.

The assumption of permissibility, together with the use of a right invariant prior, also turns out to be enough to essentially eliminate the marginalization paradoxes of Dawid et al. [54], and also some similar inconsistencies; cp. Helland [111, 112]. These paradoxes were criticized by Jaynes in a series of arguments which later appeared in the book [131]. The critique has been countered in [55]. Here I will not go into the details of this discussion, only demonstrate a point where the concept of permissibility can be used to give a cruial, constructive argument.

The main problem in [54] is a violation of the plausible:

Reduction principle
Assume that a general method of inference, applied to data (y, z), leads to an answer that in fact depends on z alone. Then the same answer should appear if the same method is applied to z alone.

A Bayesian implementation of this principle runs as follows: Assume first that the probability density $p(y, z|\eta, \zeta)$ depends on the parameter $\theta = (\eta, \zeta)$ in such a way that the marginal density $p(z|\zeta)$ only depends upon ζ. Then the following implication should hold:

Bayesian reduction principle
If (a) the marginal posterior density $\pi(\zeta|y, z)$ depends on the data (y, z) only through z, then (b) this $\pi(\zeta|z)$ should be proportional to $a(\zeta)p(z|\zeta)$ for some function $a(\zeta)$, so that it is proportional to a posterior found only from the z data.

For a proper prior $\pi(\eta, \zeta)$ this can be shown to hold with $a(\zeta)$ being the appropriate marginal prior density $\pi(\zeta)$. Dawid, Stone and Zidek [54] gave several examples where the implication above is violated by improper priors of the kind that we sometimes expect to have in objective Bayes inference.

For our purpose, the interesting case is when there is a transformation group G defined on the parameter space. Under the assumption that ζ is maximal invariant under G and making some regularity conditions, it is first shown in [54] that it necessarily follows that $p(z|\eta, \zeta)$ only depends upon ζ, next (a) is shown to hold always, and finally (b) holds if and only if the prior is of the form $\nu_G(d\eta)d\zeta$, where ν_G is the right invariant measure, and the measure $d\zeta$ is arbitrary. Thus for this situation with such a prior not only does the Bayesian reduction principle hold; we also have that the premises of the principle are automatically satisfied.

The strong assumption made above was that ζ is maximal invariant. In a second class of examples, it is shown in [54] that this assumption cannot be violated arbitrarily. In [112] the following result was shown, indicating that it is essentially enough to make the much weaker assumption that ζ is a permissible subparameter.

Theorem 3.4.1.
Assume that ζ is permissible, and let K be the subgroup of G given by $K = \{g : \zeta(\theta g) = \zeta(\theta)$ for all $\theta\}$, Then ζ is maximal invariant under K. Assume also that z is maximal invariant under the corresponding group acting on the sample space. Then using the right invariant prior under G on the sample space Θ, we have that any data (y, z) leads to a posterior of ζ proportional to the one obtained from only data z.

Since the marginalization paradox is not mentioned later in this book, I will not go further into the arguments, but continue with the discussion on permissibility, a concept which will be crucial at some points later.

In fact, on all this background it is of interest to characterize the permissible subparameters.

Proposition 3.4.1. *The subparameter η is permissible iff the level sets of the* *function $\eta = \eta(\theta)$ are transformed onto other level sets by elements of the group G. Some cases of interest are:*

a) The full parameter θ is always permissible.

b) Any invariant function ($\eta(\theta g) = \eta(\theta)$ for all g, θ) is permissible. For such functions, the action of G on η is trivial.

c) If η is permissible with range E and if γ is a one-to-one map from E onto a space F, then ζ given by $\zeta(\theta) = \gamma(\eta(\theta))$ is permissible.

d) If $\eta_1(\theta), \ldots, \eta_k(\theta)$ are k permissible functions, then the vector parameter $\psi(\theta) = (\eta_1(\theta), \ldots, \eta_k(\theta))$ is permissible, and vice versa.

e) Under the location group ($\mu \mapsto \mu + a, \sigma \mapsto \sigma$) in a location and scale family $\theta = (\mu, \sigma)$, a one-dimensional parameter $\eta = \eta(\mu, \sigma)$ is permissible iff it is either a 1-1 function of σ or if it is of the form $\eta = \psi(\mu + \phi(\sigma))$, where ψ is a 1-1 function and ϕ is an arbitrary function. In particular, the parameters μ, σ and $\mu + k\sigma$ are permissible.

f) Under the location and scale group ($\mu \mapsto b\mu + a, \sigma \mapsto b\sigma$) in a location and scale family $\theta = (\mu, \sigma)$, a one-dimensional parameter $\eta = \eta(\mu, \sigma)$ is permissible iff it is of the form $\eta = \zeta(k_1\mu + k_2\sigma)$ for some 1-1 function ζ. In particular, the parameters μ, σ and $\mu + k\sigma$ are permissible.

g) Assume that y is multivariate normal with expectation vector $K\beta$ for some fixed matrix K and with a covariance matrix Σ depending upon a set of parameters. Consider the translation group G given by $y \mapsto y + Kc$, so that the corresponding parametric group is given by $\beta \mapsto \beta + c, \Sigma \mapsto \Sigma$. Then every set of linear combinations of the form $\eta = C\beta$, where C is some fixed matrix, will be permissible. Since any function of the parameters in Σ will be invariant, it will also be permissible. This is the group associated with the restricted maximum likelihood estimator; see Section 3.9.

h) Assume that some parameter η is permissible under some group G. Let E be a set in the range of η. Then the indicator parameter $\chi_E(\eta)$, which is 1 iff $\eta \in E$, otherwise 0, is permissible iff E is equal to some union of orbits of the group G as action on η.

The simplest case of point h) is when the parameter η is invariant. Then the condition holds for any (measurable) E. Otherwise it will be difficult to meet the condition in general. This explains to a certain extent why group theoretical aspects of hypothesis testing are mainly of interest for invariant parameters, while the corresponding estimation theory is of interest for the larger class of permissible parameters.

If some given subparameter should not be permissible, one can always solve this in principle by going to a subgroup. In fact, one can easily show that there is a maximal subgroup of G with respect to which the subparameter is permissible. If this is a genuine subgroup, it cannot be transitive on Θ.

Lemma 3.4.1. *Given a subparameter η there is always a maximal subgroup G_0 of G such that η is permissible with respect to G_0.*

Proof.

Let G_0 be the set of all $g \in G$ such that for all $\theta_1, \theta_2 \in \Theta$ we have that $\eta(\theta_1) = \eta(\theta_2)$ if and only if $\eta(\theta_1 g) = \eta(\theta_2 g)$. Then G_0 contains the identity. Furthermore, using the definition with θ_1, θ_2 replaced by $\theta_1 g_1, \theta_2 g_1$, it follows that $g_1 g_2 \in G_0$ when $g_1 \in G_0$ and $g_2 \in G_0$. Using the definition with θ_1, θ_2 replaced by $\theta_1 g^{-1}, \theta_2 g^{-1}$, it is clear that it contains inverses. Hence G_0 is a group. It follows from the construction that it is maximal.

A standard concept in the statistical literature involving group invariance is the concept of *equivariant estimator* (see [146]), a concept which can be closely linked to that of a permissible parameter. Roughly speaking an estimator is equivariant if it transforms under the group in the same way as the parameter to be estimated; more precisely:

Definition 3.4.2. *Let $\eta(\cdot)$ be a permissible parameter as defined above, so that a group of transformations G is defined on η by $(\eta g)(\theta) = \eta(\theta g)$. An estimator $\hat{\eta}$ is then called* equivariant *if $\hat{\eta}(yg) = (\hat{\eta}g)(y)$ for all g, y.*

For more information on equivariant estimators, see [33, 210].

The best equivariant estimator, which will be discussed in some detail below, will in general depend upon the group used (see examples given in [146] and in [24]). Thus again the choice of group is crucial. In fact, for a group which is transitive on the parameter space, and for a quadratic loss function, the assumptions above imply that the formal Bayes estimator under right invariant prior is also the best equivariant estimator. (See Corollary 3.5.5 below.)

Another use of a specified group in at least some statistical inference problems (see also [87]) is: One will usually condition the statistical analysis upon the orbits in the sample space S; at least if the orbit index has a distribution which is independent of the parameter (which it will when the parameter group is transitive). Hence a different choice of group may often mean different conditioning. It is a well known problem in statistics that the ancillary (see Section 2.13) to condition upon may be non-unique. Specification of a group leads to a unique orbit index in the case where the parameter group is transitive.

The situation with a transitive transformation group on the parameter space is a very common one, since, because one purpose of a statistical model is to condense information, the parameter space usually is 'less' than the sample space. A typical example is the case where the distribution of a set of observations $y_1, ..., y_n$ depends upon a location parameter μ and a scale parameter σ. Then it is natural to look at the translation- and scale-group $y_i \mapsto a + by_i$ with the corresponding parameter group given by $\mu \mapsto a + b\mu$, $\sigma \mapsto b\sigma$. It is easy to see that the last transformation group is transitive, while the group in the sample space has orbits indexed by the so called configuration, for instance given by

$$a = \{\frac{y_{i+1} - y_i}{y_2 - y_1}; i = 2, ..., n - 1\}.$$

It is often claimed that all inference should be conditional, given such ancillary variables (this is related to the conditionality principle; see Section 2.13 again), and in particular that the uncertainty shall be given as conditional given a. If the observations are normally distributed, it follows from a well known theorem by Basu [146] that the mean $\overline{y}$ and the standard deviation s are independent of a, so it doesn't matter for the inference about (μ, σ) whether or not we condition with respect to a.

Here is one more statistical use of a given group, complementary to the one given above: The orbit index (a) in the sample space will be a maximal invariant under the group. Furthermore, the distribution of a only depends upon the maximal invariant (τ) in the parameter space. Useful inference on τ can therefore be performed using the marginal distribution of a, for instance computing maximum likelihood estimates from this distribution, not from the full sample distribution. The well known restricted maximum likelihood method for estimating variance components in mixed models can - as pointed out by McCullagh [155] - be seen in this perspective; see also Section 3.9 below.

3.5   Estimation under Symmetry

Summary: From a theoretical point of view this is a central section. Recall that in statistical inference the risk function is the expectation of the chosen loss function. A group G is assumed to act on the parameter space. First, in Theorem 3.5.2 a general identity is proved on a generalisation of the risk function. Then, in Subsection 3.5.2 various consequences are found for the risk function itself; in particular, in Corollary 3.5.4 conditions are given to ensure that the risk function is an overall constant. Finally, in Corollary 3.5.5 invariant quadratic loss is assumed, and under this condition and a condition on transitivity it is proved that the best equivariant estimator (Pitman estimator) is equal to the Bayesian estimator under right invariant prior.

It is well known that under quite general conditions [24, 146] risk functions are constant functions of the parameters for loss functions which are invariant under some given group; in particular this holds for natural loss functions involving equivariant estimators. This is the starting point for several strong results in the literature. I have been interested in finding (i) results in this direction that are constructive in the sense that explicit expressions can be given for the best equivariant estimators and (ii) the most general result of this kind. Partial results have among others been given in [86, 193, 127, 31, 24, 133], but both my aims seem first to be achieved by the results by Eaton [68] and Eaton and Sudderth [73]; similar results to these will be given below.

3.5.1   *The Main Result*

As before, let the sample space be S, the parameter space Θ and the model a family of probability measures $\{Q^\theta\}$ on S. In addition we need an action space - to be thought here of as the space of values of a class of estimators. Again we will disregard measurability questions.

The group G of transformations is acting upon S, and also induces a group of transformations on Θ by (3.1). In this section we also assume that the measures Q^θ are dominated by, that is, have a density with respect to, a fixed measure λ on S, which we will assume is right invariant. It is shown in Eaton [68], p. 75 that the right invariant choice here is the only one which implies that the measure behaves with respect to orbits as it should.

We then also assume that the densities

$$q^\theta(y) = \frac{dQ^\theta}{d\lambda}(y)$$

satisfy

$$q^{\theta g}(yg) = q^\theta(y) \tag{3.5}$$

for all y, θ, g. From this property it is an easy exercise, using the fact that λ is right invariant, to show that (3.1) holds, as desired.

Let us focus upon a subparameter $\eta = \eta(\theta)$. Let the loss function $L(\hat{\eta}, \theta)$ denote the loss we experience when we estimate the parameter by $\hat{\eta}$, given that the full parameter is θ, so that in particular the real value of η is $\eta(\theta)$. Usually, the loss function is positive and is convex as a function of $\hat{\eta}$, with a minimum for $\hat{\eta} = \eta(\theta)$.

Definition 3.5.1.   *The loss function is said to be invariant under G if $L(\hat{\eta}g, \theta g) = L(\hat{\eta}, \theta)$ for all $g \in G$.*

This is a very desirable property when G is associated with the statistical model.

Recall from Definition 2.6.3 that the risk function, a function of the parameter, is the expected loss under the model: $R(\theta) = E^\theta(L(\hat{\eta}(y), \theta))$.

In this and in the next section we will assume that the loss function is invariant under the group G. As a technical requirement we will also assume that the group action G is proper (for a definition see Appendix A.2.1; when we in the following talk about a proper group G, we will mean that the group actions both on S and on Θ are proper.).

Recall that λ was the right invariant measure on S with respect to which the densities of Q^θ were evaluated. Now fix a point $y \in S$. Then the measure on G introduced by $\nu_G(dg) = \lambda(d(yg))$ is right invariant. This corresponding left invariant measure $\mu_G(dg) = \Delta_G(g)\nu_G(dg)$ is used in the following basic result.

Theorem 3.5.2. *Let $\hat{\eta}(\cdot)$ be an equivariant estimator. We make the assumptions above on G, L and on the model, and assume that the group is transitive on the parameter space. Then for a fixed y on some orbit O_y in S we have*

$$\int_{O_y} L(\hat{\eta}(z), \theta) Q^\theta(dz) = \int_\Theta L(\hat{\eta}(y), \theta) q^\theta(y) \nu(d\theta). \tag{3.6}$$

<u>Proof.</u>

$$\int_{O_y} L(\hat{\eta}(z), \theta) q^\theta(z) \lambda(dz)$$

$$= \int_G L(\hat{\eta}(yg), \theta) q^\theta(yg) \lambda(d(yg))$$

$$= \int_G L(\hat{\eta}(y), \theta g^{-1}) q^{\theta g^{-1}}(y) \nu_G(dg)$$

$$= \int_G L(\hat{\eta}(y), \theta h) q^{\theta h}(y) \nu_G(dh^{-1})$$

$$= \int_G L(\hat{\eta}(y), \theta h) q^{\theta h}(y) \Delta_G(h) \nu_G(dh)$$

$$= \int_G L(\hat{\eta}(y), \theta g) q^{\theta g}(y) \mu_G(dg),$$

where we have used (3.5) and in the next to last line (A.22).

Finally we use the general result that for a *right* invariant measure ν and a transitive parameter group we have

$$\int \phi(\theta) \nu(d\theta) = \int \phi(\theta) \nu(d\theta h) = \int \phi(\theta h^{-1}) \nu(d\theta) = \int \phi(\theta h^{-1}) \nu(d(\theta h^{-1} h))$$

$$= \int \phi(\theta h^{-1}) \nu_G(dh) = \int \phi(\theta g) \nu_G(dg^{-1}) = \int \phi(\theta g) \mu_G(dg).$$

3.5.2 *Consequences*

First look at the simple situation where G is transitive both on the sample space and on the parameter space. Then the left-hand side of equation (3.6) gives the risk function, and the right-hand side is constant in θ.

But one can say more than this. In general the left-hand side of (3.6) depends only on the orbit in S; hence both sides can be regarded as a function of the orbit index a in S. In particular, the right-hand side is independent of the choice of $y = y_a$ on the orbit given by a. Summing over orbits, the left-hand side of (3.6) becomes the risk, so:

Corollary 3.5.3. *Assume that the group G is proper and that the loss function L is invariant under G. Then there are positive constants c_a for each orbit O_a in the sample space such that the risk function is*

$$R(\theta) = \sum_a c_a \int_G L(\hat{\eta}(y_a), \theta g) q^{\theta g}(y_a) \mu_G(dg) \tag{3.7}$$

for one fixed left invariant measure μ_G, where each y_a is an arbitrarily chosen point on the orbit O_a. If there is a continuum of orbits, the sum is replaced by an integral.

Proof.
Sum both sides of (3.6). On the right-hand side of that equation, the left invariant measure μ_G depends on the orbit in S. But since all left invariant measures on G are proportional, the result follows. The value of c_a depends upon the normalization of the right invariant measure on each orbit.

Corollary 3.5.4. *Assume that the group G is proper and that the loss function L is invariant under G. Then the risk function is constant on orbits in Θ. In particular, if the group action on Θ is transitive, then $R = R(\theta)$ is constant for all θ.*

Proof.
Use (3.7).

One strong point of the equation (3.6) is that the right-hand side can be used to find the best equivariant estimator explicitly, as demonstrated in several of the references mentioned above. In [127] the last integral was interpreted as an expectation with respect to a 'fiducial' distribution, but also under certain conditions as an integral with respect to a posterior distribution. If L is a quadratic loss function, it is easy to see from (3.6) that the best equivariant estimator (Pitman estimator) can be interpreted formally as the Bayes estimator under an invariant (right invariant) prior.

Corollary 3.5.5. *Let $L(\hat{\eta}(y), \theta) = B(y)\|\hat{\eta}(y) - \eta(\theta)\|^2$, and assume that this loss function is invariant under G, a proper group. Let the group transformation on the parameter space be transitive with right invariant measure ν. Then the best equivariant estimator for η is given by*

$$\hat{\eta}(y) = \int \eta(\theta)\frac{q^\theta(y)}{q(y)}\nu(d\theta),\tag{3.8}$$

with $q(y) = \int q^\theta(y)\nu(d\theta)$. This assumes that the above expression is finite.

Proof.
First fix $y \in S$. Then this y will belong to some orbit O_a. Using the fact that the right-hand side of (3.6) is independent of the choice of y in O_a, in (3.6) and independent of θ from the right-hand side expression, we expand this right-hand side to form a perfect square, and find

$$B(y) \left\| \hat{\eta}(y) - \frac{\int \eta(\theta)q^\theta(y)\nu(d\theta)}{\int q^\theta(y)\nu(d\theta)} \right\|^2 \int q^\theta(y)\nu(d\theta) + const.(y).$$

This gives an optimal choice for $\hat{\eta}(y)$ for y in each orbit in S, a choice which has the same form for all orbits of the sample space. But then this choice also must be optimal with respect to the sum over orbits of the left-hand side of (3.6), i.e., with respect to the risk function. For the case where the group action is transitive on the parameter space Θ, the optimal choice of $\hat{\eta}(y)$ will be as in (3.8).

Note that the vanishing of the cross terms above requires $\int \phi(\theta g)\mu_G(dg) = \int \phi(\theta)\nu(d\theta)$ for several different functions ϕ. This is consistent with requiring that ν is a *right* invariant measure:

$$\int \phi(\theta)\nu(d\theta) = \int \phi(\theta)\nu(d\theta h) = \int \phi(\theta h^{-1})\nu(d\theta)$$
$$= \int \phi(\theta h^{-1})\nu_G(dh) \int \phi(\theta g)\nu_G(dg^{-1}) = \int \phi(\theta g)\mu_G(dg).$$

Note that the density integrated over in (3.8) is just the posterior density $\nu(\theta|y)$ of equation (2.34) in Chapter 2. Hence the best equivariant estimator under quadratic loss when the group is transitive on the parameter space is just the Bayes estimator obtained by using the right invariant measure as a prior. This estimator is often called the Pitman estimator [175].

Example 3.5.1. In the examples 2.3.2, 2.3.6 and 2.6.2 we gave different methods for estimating the parameters α and β in the model with uniform distribution for each of the n independent observations $y_1, \ldots, y_n$, i.e., such that each y_i has a constant density $1/(\beta - \alpha)$ in the interval $\alpha < y_i < \beta$. In this example we give another Bayes estimator corresponding to a natural symmetry group associated with the model.

Look upon $\delta = \beta - \alpha$ as a scale parameter and $\mu = (\alpha + \beta)/2$ as a location parameter. The right invariant measure corresponding to the location- and scale group is $\delta^{-1}d\delta d\mu$. Let $m = \min_i(y_i)$ and $M = \max_i(y_i)$. Then we must have the necessary and sufficient inequalities $\delta > M - m$, furthermore $\mu + \delta/2 > M$ and $\mu - \delta/2 < m$, which implies $M - \delta/2 < \mu < m + \delta/2$. The joint posterior density in the range given by these inequalities is:

$$\nu(\delta, \mu|y) = \frac{\delta^{-n-1}}{\int_{\delta=M-m}^{\infty} \int_{\mu=M-\frac{1}{2}\delta}^{m+\frac{1}{2}\delta} \delta^{-n-1}d\mu d\delta} = \frac{n(n-1)(M-m)^{n-1}}{\delta^{n+1}}.$$

By integrating this, we find that the marginal posterior for δ is

$$\nu(\delta|y) = \frac{n(n-1)(M-m)^{n-1}(\delta - (M-m))}{\delta^{n+1}},$$

and the Bayes estimator, the expectation under this density, is:

$$\hat{\delta} = \frac{n}{n-2}(M-m).$$

The marginal posterior density for μ is symmetrical around $(m+M)/2$, so $\hat{\mu} = (m+M)/2$. This gives

$$\hat{\alpha} = \hat{\mu} - \frac{1}{2}\hat{\delta} = \min_i(y_i) - \frac{1}{n-2}(\max_i(y_i) - \min_i(y_i)), \tag{3.9}$$

$$\hat{\beta} = \hat{\mu} + \frac{1}{2}\hat{\delta} = \max_i(y_i) + \frac{1}{n-2}(\max_i(y_i) - \min_i(y_i)), \qquad (3.10)$$

thus only slightly different results from what we fund in Example 2.6.2. But note that the estimators (3.9, 3.10) have an optimality property also for finite n: They are derived from the best equivariant estimators (under quadratic loss) under the location- and scale- group.

For completeness it should be mentioned that the unbiased estimators of this type are slightly different again, namely

$$\hat{\alpha} = \min_i(y_i) - \frac{1}{n-1}(\max_i(y_i) - \min_i(y_i)),$$

$$\hat{\beta} = \max_i(y_i) + \frac{1}{n-1}(\max_i(y_i) - \min_i(y_i)).$$

(The distribution of M is found from $P(M \leq z) = \prod_i P(y_i \leq z)$. It is easiest to start with the case $\alpha = 0$, $\beta = 1$.)

It is very interesting to note that, when G is not transitive on Θ, (3.9) implies an estimator corresponding to each orbit there. Minimizing the expression (3.9) leads to a unique estimator corresponding to each orbit in the parameter space:

$$\hat{\eta}_\tau(y) = \frac{\int_G \eta(\theta g) q^{\theta g}(y) \mu_G(dg)}{\int_G q^{\theta g}(y) \mu_G(dg)}. \qquad (3.11)$$

Theorem 3.5.6. *For each orbit τ in the parameter space, the estimator (3.11) is equivariant and minimizes the risk $R(\theta)$ corresponding to the quadratic loss function in Corollary 3.5.5 for θ belonging to the orbit τ.*

The proof of Theorem 3.5.6 is deferred to Appendix A.2.4.

Theorem 3.5.6 will later become our main argument for the model reduction policy of only allowing model reductions to an orbit or to a set of orbits on the parameter space. Further reductions can never lead to an improvement of the data analysis: We already have an optimal estimator on orbits; see Sections 3.7 and 3.8 below.

3.6   Credibility Sets and Confidence Sets

Summary: In this section I prove the following: If the group is transitive both on the parameter space and on the sample space, then confidence intervals and credibility intervals with the same probability are numerically equal.

Essentially as in [24] and in other books on Bayesian statistics, define a $100(1 - \alpha)\%$ credibility set as a set $C(y)$ in the parameter space whose posterior probability given data y is $1 - \alpha$. We will concentrate on the non-informative right invariant prior ν, so that the posterior is $(q^\theta(y)/q(y)) \cdot \nu(d\theta)$, where $q(y) = \int q^\theta(y)\nu(d\theta)$.

Definition 3.6.1. *The credibility set is defined by*

$$\int_{C(y)} \frac{q^\theta(y)}{q(y)} \nu(d\theta) = 1 - \alpha. \tag{3.12}$$

A confidence set $C(y)$ is also a set in the parameter space, depending upon data y, but the probability interpretation is completely different: $Q^\theta(\theta \in C(y)) = 1 - \alpha$, where the probability is over the distribution of y. The link between the two concepts, however, is easily found from Theorem 3.5.2, using $L(y, \theta) = 1 - I(\theta \in C(y))$.

Theorem 3.6.1. *Let G be proper and acting transitively both on S and on Θ. Assume that the collection of sets $\{C(y)\}$ satisfies the transformation law $C(yg) = (Cg)(y)$ for all y and g. Then each $C(y)$ is a credibility set if and only if it is a confidence set, and the two probabilities associated with the sets are the same.*

In fact, Theorem 3.5.2 gives a stronger statement: The corresponding conditional probabilities, conditioned upon the orbits in S, are equal. We will still assume that the group acts transitively on the parameter space, and we will take a right invariant prior on that space.

Corollary 3.6.2. *Let $\eta(\theta)$ be a one-dimensional continuous permissible parametric function, and let $\hat{\eta}_1(y)$ and $\hat{\eta}_2(y)$ be two equivariant estimators under a group G which is proper and transitive on the parameter space. Define $C(y) = \{\theta : \hat{\eta}_1(y) \leq \eta(\theta) \leq \hat{\eta}_2(y)\}$. Then $C(y)$ is a credibility set and a confidence set with the same associated probability/confidence level.*

Proof.
Since the mapping g defined by $\eta(\theta g) = (\eta g)(\theta)$ is a continuous 1-1 mapping from a one-dimensional connected set onto another one-dimensional set, it must preserve or reverse ordering. Without loss of generality, extend the definition of $C(y)$ to $\{\theta : \hat{\eta}_1(y) \leq \eta(\theta) \leq \hat{\eta}_2(y)\} \cup \{\theta : \hat{\eta}_2(y) \leq \eta(\theta) \leq \hat{\eta}_1(y)\}$. One of these components must be empty. So $(Cg)(y) = \{\theta g : \hat{\eta}_1(y) \leq \eta(\theta) \leq \hat{\eta}_2(y)\} \cup \{\cdots\} = \{\theta : \hat{\eta}_1(y) \leq \eta(\theta g^{-1}) \leq \hat{\eta}_2(y)\} \cup \{\cdots\} = \{\theta : (\hat{\eta}_1 g)(y) \leq \eta(\theta) \leq (\hat{\eta}_2 g)(y)\} \cup \{\cdots\} = C(yg)$. Hence the result follows from Theorem 3.6.1.

A simple example is the following: Let $y_1, \ldots, y_n$ be independent and identically distributed normally distributed observations with sample mean $\bar{y}$ and sample standard deviation s. Let k be chosen so that the confidence statement $\bar{y} - ks \leq \mu \leq \bar{y} + ks$ has confidence coefficient $1 - \alpha$. Then the interval can also be given a definite probability interpretation: $1 - \alpha$ is also equal to the posterior probability of the interval when the prior is right invariant measure under the translation- and scale-group. Note that the action of G is transitive here. Similar examples can be found in a lot of cases where a non-informative prior is used in Bayesian analysis; see for instance [176].

The results of this section also have immediate consequences for *confidence distributions*, an area which has been discussed recently by Schweder and Hjort [182] as a frequentist alternative to Bayes posteriors. Briefly, if $[\eta \leq \eta_\beta(y)]$ is a one-sided confidence set for the parameter η with confidence coefficient β, and this is calculated for all β, the functional relation $F(\eta_\beta(y)) = \beta$ is equivalent to some $F(\eta_0) = \beta_0(y)$, which can be looked upon formally as giving a 'distribution' of η for fixed data, the confidence distribution of the parameter. A question of interest is when this is equal to a Bayesian posterior for some prior. The following immediate consequence of Corollary 3.6.2 gives a partial answer:

Corollary 3.6.3. *Assume that the statistical model is invariant under a group G, and that η is a one-dimensional continuous permissible parametric function of the model parameter θ. Assume that the group G is transitive on the parameter space. Then for fixed orbit index (ancillary) a under G the confidence distribution for η will be equal to the posterior distribution under right invariant prior.*

Using Theorem 3.5.2 it is possible to generalize this result to the multiparameter case.

3.7   Examples. Orbits and Model Reduction

Summary: *The general model reduction philosophy: Model reduction should always be to a set of orbits of the relevant group acting on the parameter space, is formulated and illustrated by examples.*

I will not attempt any general theory of model reduction here, but the following remark seems rather obvious from the preceeding discussion: For each orbit of the group acting upon the parameter space, the Pitman estimator of Theorem 3.5.6 gives a good solution of any estimation problem within each orbit. Hence, if the purpose of a model reduction should be to be able to obtain more precise estimates from the model, there seems to be little reason to reduce the model further within orbits. In this book I will largely stick to the following

Model reduction policy. *When a group G is associated with a statistical model, every model reduction should be to an orbit or to a set of orbits of G as acting upon the parameter space Θ.*

Such a model reduction may first be thought of as a direct restriction of the space Θ, but from our point of view it is better to look upon it, at least partly, as going from the space Θ to a new space Λ through a specific function $\lambda(\theta)$.

Definition 3.7.1. *Let T be the original set of orbits of Θ. For each orbit τ in T, pick a fixed point θ_τ, and let the points of that orbit have a representation $\theta = \theta_\tau g_\theta$. (If the stability group of θ_τ is trivial, then the choice of $\{g_\theta\}$ is unique; otherwise it must be selected.)*

Let T_1 be a set of orbits on which we want to reduce the model. Then the reducing function λ can be defined as follows: First define a function ζ from T onto T_1, and then let $\lambda(\theta) = \theta_{\zeta(\tau)} g_\theta$.

One point of the above construction is that the function λ will be permissible, which can be verified directly. This will also ensure that the reduced parameter space is invariant under the group, which we from the beginning of this Chapter have held up as an essential requirement for any model. Thus our model reduction policy ensures these basic requirements, which may be considered an argument for the policy. Further arguments are given by the examples below. It must be emphasized, though, that not every imaginable model reduction satisfying this policy is reasonable. Further arguments in the form of an improved prediction error or in terms of providing a value for parameters in situations were parameters did not have a value before, must be sought.

Model reduction is thought of in applied statistics as a means to perform approximate inference. But, as also discussed in Section 2.15, in certain cases parts of the parameter do not have a value relative to the experiment. Then the model reduction may lead to an exactly true model. This is the situation we will meet in our approach to quantum mechanics.

In modern applied statistics, on the other hand, a common attitude seems to be that all models are approximations (see [37]), and that model reduction is seen as a bias versus variance issue.

These two situations seem different, but as discussed in Section 2.15 by pointing at the collinearity issue in multiple regression, they must also be considered to be related.

Some examples which give additional support to the model reduction policy above, are:

Example 3.7.1. Look at two independent samples, $x_1, ..., x_m$, which are independently normal (μ_1, σ_1^2) and $y_1, ..., y_n$, which are independently normal (μ_2, σ_2^2). Use the translation- and scale-group given by $\mu_1 \mapsto a_1 + b\mu_1$, $\sigma_1 \mapsto b\sigma_1$, $\mu_2 \mapsto a_2 + b\mu_2$, $\sigma_2 \mapsto b\sigma_2$. (A common b must be used in order that $\mu_1 - \mu_2$ shall be permissible; or more directly: It is meaningless to test a hypothesis $\mu_1 = \mu_2$ if the two samples are transformed into different units.)

Then the orbits of the group in the parameter space are given by $\sigma_1/\sigma_2 =$ *constant*. A very common model reduction is given by $\sigma_1 = \sigma_2$.

Example 3.7.2. Consider a two way analysis of variance with expectations $\mu + \alpha_i + \beta_j + \gamma_{ij}$, and a group generated by all permutations of the index i and by all permutations of the index j. Then an obvious reduced model is given by the orbit where the expectation is $\mu + \alpha_i + \beta_j$.

Example 3.7.3. In a multiple regression it is not uncommon that all explanatory variables x_j are measured in different units. Then a natural group in the sample space (permitting inclusion of the x-variables in this space) is given by separate scale changes $x_j \mapsto k_j x_j$ $(j = 1, 2, ...)$. This induces a similar group on the regression parameters, namely $\beta_j \mapsto k_j^{-1}\beta_j$, and all orbits in the parameter space are given by putting some of the β_j's equal to 0. These reduced models are well-known from many applications of regression analysis, and criteria like C_p or AIC are used to discriminate between them.

Example 3.7.4. Assume that for some sample of independent, identically distributed observations you start by modelling it using a model depending upon many parameters. Let the usual translation- and scale-group act on this large parameter space. Then one orbit is given by the $N(\mu, \sigma^2)$ distribution, a not uncommon model reduction.

Example 3.7.5. Consider the rotation group for a multivariate data set. This induces the transformations $\Sigma \mapsto C'\Sigma C$ for the covariance matrix Σ, where C is some orthogonal matrix. This is an extremely non-transitive group with orbits equal to every set of eigenvalues of Σ, counting multiple eigenvalues with their multiplicity. It is difficult to imagine a situation where it is of interest to take a single orbit as a reduced model, but sets of orbits can make interesting reduced models, say those where the number of different eigenvalues is some fixed number, or those where the 5 smallest eigenvalues are equal.

3.8   Model Reduction for Subparameter Estimation and Prediction

Summary: The model reduction philosophy of the previous section is discussed for the case of a subparameter and illustrated for a particularly important case: That of rotationally invariance among the regressors in a random x multiple regression, leading to the chemometrician's partial least squares regression. This issue will be taken up again in Chapter 8.

The model reduction policy of Section 3.7 can lead to many different situations. The extreme situations are: First consider the case where G is transitive on Θ.

Then there is no possibility of a model reduction according to our policy. Secondly, we have cases with very many orbits, like in Example 3.7.5. This situation is also undesirable as it stands, for it leaves you with so many options that at least at first sight there is no possibility of selecting a good model reduction, say by some cross validation procedure.

3.8.1 *Estimation of Subparameters*

Fortunately, in most such cases we are not interested in the full parameter θ, but in some subparameter η. In this section we will assume that this parameter η is permissible. (By Lemma 3.4.1, this can always be achieved by replacing G by a subgroup, a procedure which admittedly may increase the number of orbits of the group action on Θ further.)

For a permissible subparameter η, the group G will act on its range by $(\eta g)(\theta) = \eta(\theta g)$. The number of orbits of G as acting upon the range of η will typically be smaller than the number of orbits on Θ, often considerably smaller. Often the action on η may even be transitive. Then results similar to (3.8) may also be obtained. We will first consider this equation as a formula for an estimator, where ν is an arbitrary right invariant prior. Recall that a right invariant prior is unique within orbits, but that an arbitrary measure on the orbit index must be supplied.

The simplest case is when $\theta = (\eta, \zeta)$ actually is linked to η in such a way that the remaining parameter ζ also is transitive. Then there is a unique right invariant measure corresponding to this parameter, and the integral over ζ can be carried out independently on the right-hand side of (3.8). This equation then gives a unique solution for $\hat{\eta}$, which will be the best equivariant estimator.

A more exact discussion can be provided by taking the equation (3.9) as a point of departure. Minimizing the expression here leads to a unique estimator corresponding to each orbit in the parameter space, as pointed out in Theorem 3.5.6:

$$\hat{\eta}_\tau(y) = \frac{\int_G \eta(\theta g) q^{\theta g}(y) \mu_G(dg)}{\int_G q^{\theta g}(y) \mu_G(dg)}. \tag{3.13}$$

Depending upon the type of orbit, the estimators (3.13) can be essential to the problem at hand, but they can also be quite trivial: For any orbit τ which is contained in some set of the form $\{\theta : \eta(\theta) = \eta_0\}$ the estimator $\hat{\eta}_\tau$ will only take the trivial value η_0, so estimators corresponding to such orbits can immediately be discarded.

A similar situation occurs if the numerator and the denominator of (3.13) can take the same value for several orbits, and the estimator may then be taken to correspond to the union of these orbits.

Typically, we will have this situation if for some part of the orbit index τ we have that $\eta(\theta)$ is independent of this part, but it is only contained in $q^\theta(y)$. A simple example might be if we want to estimate the norm of the expectation vector of a multivariate data set under rotational symmetry. Then one can adjoin the integral over this orbit index part to the denominator and the numerator of (3.13), but this integral will then cancel if the remaining integral takes the same value for all orbits corresponding to this part of the orbit index.

3.8.2   *Multiple Regression under Rotation Invariance*

To illustrate some of these ideas, we first turn to the case of regression analysis with random regressors. This will ultimately lead to the population partial least squares regression model which was introduced in Helland [105], and also discussed in other publications. This model was argued there to correspond to the partial least squares procedure for biased regression, introduced by chemometricians through a particular algorithm. This correspondence will be further discussed and developed in Chapter 8.

Let $(x_i, y_i; i = 1, \ldots, n)$ be n independent observations, where the vectors x_i are p-dimensional, are centered and have components with the same units so that rotation of this vector is a meaningful operation. Let $\Sigma = V(x_i)$, $\sigma^2 = Var(y_i|x_i)$, and the β be the regression vector of y_i on x_i, assumed linear, so that we have a linear model, conditioned upon $\{x_i\}$ of the form

$$y = X\beta + e, \qquad (3.14)$$

where (X, y) has rows $(x_i', y_i; i = 1, \ldots, n)$ and e is a vector of independent errors with expectation 0 and variance σ^2. Such a linear conditional model will hold if the data are multinormal, but also in other cases.

Let C denote rotation matrices, and let the rotation group act upon these data: $(x_i, y_i) \mapsto (Cx_i, y_i)$, so that the parameter transformation is given by $(\Sigma, \beta, \sigma) \mapsto (C\Sigma C', C\beta, \sigma)$. In Example 3.7.5 it was pointed out that the orbits when concentrating on the parameter matrix Σ are given by the ordered set of eigenvalues of that matrix, where multiple eigenvalues count by their multiplicity.

In Helland [109] it was shown that orbits of the full parameter in this case are given as follows.

Theorem 3.8.1. *Let the spectral decomposition of Σ be given by*

$$\Sigma = \sum_{k=1}^{q} \lambda_k P_k, \qquad (3.15)$$

where $\lambda_1 > \lambda_2 > \ldots > \lambda_q$ and $P_k; k = 1, \ldots, q$ are projection matrices upon spaces V_k of dimension v_k. Then the orbits of $\theta = (\Sigma, \beta, \sigma)$ are given by constant values of all of

(1) *q;*

(2) *the relative orientation of the ordered set of spaces V_k, i.e., the signs of the cosinus of the pairwise angles between these spaces;*

(3) *$v_1, \ldots, v_q$;*

(4) $\lambda_1, \ldots, \lambda_q$;

(5) $\gamma_1, \ldots, \gamma_q$, *where* $\gamma_k = \| P_k \beta \|$ *is the absolute value of the projection of the regression vector upon the vector space corresponding to the eigenvalue* λ_k *of* Σ;

(6) σ.

The theorem is proved in Appendix A.2.3.

Assume now that we, in agreement with what is common in regression, have that the parameter of interest is not the full set θ, but only the regression parameter β. In terms of the orbit indices this can be written

$$\beta = \sum_{i=1}^{q} \gamma_i e_i, \tag{3.16}$$

where the unit vectors e_i are orthogonal with one vector in each of the spaces V_k upon which the orthogonal projection operators P_k project, and the γ_i are non-negative numbers. In the estimation we can collect together the orbits corresponding to different values of $\lambda_1, \ldots \lambda_q, v_1, \ldots, v_q$, as these give the same value of β and the same conditional model density.

But still there are many orbits to choose from: Each set of values of the vector $(\gamma_1, \ldots, \gamma_q, \sigma)$ will correspond to one orbit (collected together as above). Since the e_i's occur in a symmetric way in (3.16), the ordering of these is irrelevant. Thus the ordering of the γ_i's is arbitrary. This means that we have a further reduction of the number of orbits in this model: Vectors $(\gamma_1, \ldots, \gamma_q)$ whose components are permutations of each other, may be taken as belonging to the same orbit.

Still this leaves us with too many orbits if we want to select a model reduction to a set of orbits through a simple criterion like cross validation.

3.8.3 *Towards Partial Least Squares Regression*

The fact that we have too many orbits to choose from, is an indication that we have taken a too small group as a point of departure. Now we augment the rotation group on Θ by adding the transformations g given by

$$(\gamma_1, \ldots, \gamma_q)g = (c_1 \gamma_1, \ldots, c_q \gamma_q), \tag{3.17}$$

for positive c_i, that is, an independent scale transformation of each γ_i. This turns out to be exactly what we need to arrive at a sitation where model selection can be done by simple cross validation, and at the same time what we need to reproduce the population model of partial least squares [105] by model reduction.

One could in principle also include a scale transformation of σ, but this turns out to give a too large group. Specifically, with this group, the best equivariant estimator turns out to be equal to the prior mean, so the data are not used at all. Hence we fix σ at some value, later to be estimated separately.

Theorem 3.8.2. *Consider the random regression model formulated above, and on the parameter space Θ of this model define the group G composed from the rotations in the x-space together with the independent scale transformations of each γ_i. Assume that we are interested in the best equivariant estimator of β as given by (3.13) for each orbit of the group, so that orbits which give the same value for this formula are joined together.*

(a) The orbits of this group are characterized by the single number m, the number of non-zero parameters γ_i.

(b) Alternatively, m is given by dimension of the space spanned by the Krylov sequence $\Sigma\beta, \Sigma^2\beta, \ldots$.

(c) Alternatively again, in the population PLS algorithm of Section 8.2 below, m is the first number which give weights $\omega_{m+1} = 0$.

(d) Finally, m is the number of scores τ_k and the number of loadings π_k in the same population PLS algorithm.

A major point now is that a single number m can be easily determined by crossvalidation or by using some test variable.

<u>Proof.</u>

(a) For vectors $(\gamma_1, \ldots, \gamma_q)$ with non-negative components under independent scale transformation of the components, we first remark that any non-zero value of a component may be transformed to any other non-zero value. But 0 is only transformed to 0. Hence the orbits are characterized by the numbers of and the positions of the non-zero components in the vector. In addition, we remarked at the end of the previous subsection that permutation of the components may be taken as a part of the group. Therefore, the orbits of the full group are given just by the number of non-zero components in the vector.

(b) Since $\Sigma = \sum_{k=1}^{q} \lambda_k P_k$, we have that

$$\sum_{i=1}^{a} c_i \Sigma^i \beta = \sum_{k=1}^{q} \sum_{i=1}^{a} c_i (\lambda_k)^i P_k \beta,$$

and this is 0 if and only if

$$\sum_{i=1}^{a} c_i (\lambda_k)^i = 0 \text{ for all } k \text{ such that } \gamma_k \neq 0.$$

Let m be the number of different such k, and look at this system of equations for $a = m$. The determinant coresponding to this set of equations will be a Vandermonde determinant (also called an alternant), and this determinant is non-zero since the λ_k are different. This implies that $\Sigma\beta, \Sigma^2\beta, \ldots, \Sigma^m\beta$ are linearly independent, and that dependence is introduced by adding $\Sigma^{m+1}\beta$ to this set.

For the proof of (c) and (d) we refer to Chapter 8 below, where the concepts involved will also be further discussed.

The treatment in Chapter 8 will not only give a way to characterize the value of the number m; it also gives several characterizations of the reduced parameter space, in effect the orbit corresponding to m. The right invariant measure on the parameter space corresponding to the group described above, is also easy to find: Rotation corresponds to a uniform distribution on the unit sphere of the orthogonal set of vectors $(e_1, \ldots, e_m)$. The scale transformations induce independent measures $d\gamma_1/\gamma_1, \ldots d\gamma_m/\gamma_m$. Hence it should in principle be possible to find an estimator $\hat{\beta}$ from the formula corresponding to (3.13) for each orbit, that is, each m. By the theory above this estimator should be best equivariant, that is, possess an optimality property with respect to all other reasonable estimators, in particular to the chemometricians' partial least squares estimator for each given m, then also over different m chosen by cross validation.

3.9 Estimation of the Maximally Invariant Parameter: REML

Summary: It has been proved in Lemma 3.2.3 that the orbit index in the sample space has a distribution which only depends upon the orbit index in the parameter space. Using maximum likelihood estimation from this point of departure, the REML estimator of variance components in mixed models results.

Maximum likelihood is the default estimation method in most statistical applications, even though it is well known that it can be motivated properly only for large data sets, specifically when the number n of data points is large compared to the number p of parameters in the model. If this condition does not hold, we have basically two general alternative methods of estimation [146]. The first alternative is to try to find the best unbiased estimator, most commonly through conditioning and the Rao-Blackwell theorem. The second alternative, which is treated throughout this Chapter, is to use group symmetry and seek the best equivariant estimator.

3.9.1 *On Orbit Indices and on REML*

The simplest situation (Section 3.5) is when the underlying group G is transitive on the parameter space. Then the risk is constant, and the best equivariant estimator can be found as the Bayes estimator under the right invariant prior with respect to G as prior.

When G is not transitive, a similar uniqueness property holds on the orbits of the group (Section 3.8). Recall the following facts:

1. The orbit index τ for a non-transitive group in the parameter space is the same as the maximal invariant under G. (Section 3.1.)

2. A similar statement holds in the sample space. The orbit index a here has a distribution which only depends upon τ. (Lemma 3.2.1.)

3. The risk function is constant on orbits in the parameter space. (Section 3.5.)

As shown in Theorem 3.5.5 the best equivariant estimator in the transitive case is the Bayes estimator with right invariant prior. In the non-transitive case it follows from Theorem 3.5.6 that optimizing the risk for parameters within each orbit gives a similar estimator.

But after this, the orbit index (maximal invariant) τ must be estimated in other ways. The most natural solution is to use maximum likelihood, but instead of using the full data y, restrict oneself to the sample maximal invariant $a(y)$, which is known to have a distribution which only depends on τ.

In this section we will illustrate this procedure for a special group defined on a linear mixed model, and show that to the so-called restricted (or residual) maximum likelihood (REML) estimator for the dispersion parameters of this model, and also show that this gives a different solution than the ordinary maximum likelihood estimator.

The REML estimator was first proposed (for balanced data) by W.A. Thompson [196], and then it was independently proposed and applied for unbalanced data by Patterson and Thompson [170]. After having competed with some other variance component estimators for several years, it is now a standard procedure, with a considerable attached literature.

The results below are not original; for instance they were hinted at in [146], and [155] gives an explicit, more abstract treatment. On the other hand, none of the references in the encyclopedia article [190] seem to use a group-theoretical perspective. A completely different motivation for REML is given in [188], where further references to the question of motivation are given.

3.9.2   *The Model and the Group*

Consider now an n-dimensional observation vector y which is assumed to be multi-normal with expectation $X\beta$, that is,

$$y \sim N(X\beta, \Sigma), \tag{3.18}$$

where X is a fixed $n \times p$ matrix and β is a p-vector of fixed effect parameters, and where y has a covariance matrix Σ, at least partly unknown. Explicitly, we assume that Σ depends upon a parameter vector τ, which varies over some open set, and where otherwise conditions [146] for the existence and uniqueness of solutions of the likelihood equations corresponding to the distributions given below, are satisfied.

An important special case of this model is given by the mixed model

$$y = X\beta + \sum_{k=1}^{r} Z_k u_k, \tag{3.19}$$

where the matrices Z_k have dimensions $n \times q_k$ and rank q_k, and where the u_k's are independent and multinormally distributed with expectation 0 and covariance matrix $\sigma_k^2 I$.

The parameters of the model are $\theta = (\beta, \tau)$, and in the mixed model case $\tau = (\sigma_1^2, \ldots, \sigma_r^2)$, where these variances are called variance components.

The expectation part of this general model simply means that $E(y)$ is assumed to belong to the known vector space $V = \text{span}(X)$. A natural symmetry group attached to the model is therefore the group G of translations in this space. As a first observation, if $y \mapsto yg = y + c$ for some vector $c \in V$, and if the model holds for y, then it also holds for yg. Hence the model is invariant under the group as acting upon the sample space.

The induced group on the parameter space is given by $(\beta, \tau) \mapsto (\beta + b, \tau)$, where $c = Xb$. Note that c runs through V if and only if b runs through all of $\mathbf{R}^p$. The model is clearly invariant under the transformations of the parameter space.

Obviously, the group as acting upon the parameter space is not transitive, and the orbits are just indexed by τ.

3.9.3 Estimation

For a fixed orbit, we have that $\Sigma = \Sigma(\tau)$ is fixed, and the maximum likelihood estimate is

$$\hat{\beta} = (X'\Sigma^{-1}X)^{-1}X'\Sigma^{-1}y. \tag{3.20}$$

Extending the discussion of [146], Section 3.4, we can also show that this estimator gives the minimum risk equivariant estimator of linear combinations $\phi = a'\beta$ for convex and even loss functions. This is also called the weighted least squares estimator; when variances differ it can be argued for as a general principle to use inverse variances as weights. The only difficulty with $\hat{\beta}$ is that it depends upon the unknown parameter τ, the orbit index.

So we have to estimate τ. The way to do this, relies on a property of groups given in Lemma 3.2.3: The maximal invariant a in the sample space has a distribution which depends only upon the orbit index (maximal invariant) τ in the parameter space. Hence this gives a natural setting for estimation.

Theorem 3.9.1. *(a) In the model (3.18), under the group described there, the orbit index in the sample space can be expressed as*

$$a = (I - X(X'X)^{-1}X')y.$$

(b) Let A be an $n \times (n - p)$ matrix of full rank $n - p$ such that $A'X = 0$. Then an equivalent orbit index is given by $z = A'a = A'y$. This variable z will have a non-singular distribution.

(c) The maximum likelihood estimator of τ found from the distribution of z is independent of the choice of the matrix A with the stated properties.

Remark.

The maximum likelihood estimator referred to in Theorem 3.9.1 will be the REML estimator for models of the form (3.18), in particular for mixed linear models.

It is obvious from the setting that this estimator will have many of the ordinary properties of maximum likelihood estimators. It should also be quite clear that as a principle of estimation this can be generalised to many other situations where a natural group can be associated to the statistical model.

The proof of Theorem 3.9.1 is given in Appendix A.2.3.

In a general estimation procedure, it may be impractical to find an unspecified matrix A whose columns span the space orthogonal to $V = \text{span}(X)$, so practical computation algorithms use other techniques. One of the early papers on computation in REML is [47], while a modern survey and many references can be found in [190]. The methods are much used by animal breeders, and this community has done an important job in constructing efficient programs for REML estimation in large data sets.

Example 3.9.1. Here is the simplest possible example: Let $y_1, \ldots, y_n$ be independent normal (μ, σ^2). The REML estimator of σ^2 is, according to the recipe given above, found as follows: Take $1 = (1, \ldots, 1)'$, and let A be any $n \times (n-1)$ matrix of full rank satisfying $A'1 = 0$. Then a simple calculation gives that the maximum likelihood estimator from $z = A'(y - \bar{y}1) = A'y$ is given by

$$\hat{\sigma}^2 = \frac{1}{n-1} z'(A'A)^{-1} z = \frac{1}{n-1} y' A(A'A)^{-1} A'y.$$

Here it also can be seen directly that the estimator is independent of the choice of A, since the resulting projection equals

$$A(A'A)^{-1}A' = I - 11'/n.$$

Inserting this gives the ordinary variance estimator, which is unbiased, or. more important, in an analysis of variance context, has a denominator with the correct degrees of freedom.

Example 3.9.2. Consider now a possibly unbalanced one-way analysis of variance situation, i.e., k independent groups, where group j contains n_j independent observations, each normal (μ_j, σ^2). The REML estimator for σ^2 can here be derived from a simple extension of the result of the previous example, and will be

$$\hat{\sigma}^2 = \frac{1}{\sum n_j - k} \sum_j \sum_i (y_{ij} - \bar{y}_{j\cdot})^2.$$

The corresponding maximum likelihood estimator is biased, and is found by deleting the $' - k'$ in the denominator here. The bias can be considerable if k is large and the number of observations in the groups are small. For instance, if $n_1 = \ldots = n_k = 2$, then the denominator in the REML is k, as it should be, while the maximum likelihood estimator is too small, with a denominator $2k$. A similar example from block experiments can be traced back to Neyman and Scott [165] from 1948.

The REML principle has turned out to be very useful, for instance in animal breeding, but also in other cases where linear mixed models are used. In the 1970's REML was simply one of a number of methods of estimating dispersion parameters, but now it is considered by many to be the preferred method. By what we have hoped to have illustrated, REML estimation can be regarded as an instance of a general symmetry based estimation principle which has the potential for several further applications.

3.10   Design of Experiments Situations

Summary: A brief general description of a design of experiment situation is given and fitted to the situation with a group attached to a model. This issue will be taken up again in Chapter 7.

Recall the following general description of an experiment from Section 2.14:

Consider a set Z of potential experimental units for some experiment; this set can be finite or infinite, and one may even consider an uncountable number of units. For each given $z \in Z$, let y_z be some potential response variable, and let μ_z be the expectation of y_z for the case where no treatment is introduced. One may also have a set T of potential treatments which can be applied to each unit. Let μ_{tz} be the expectation of y_z, given z, when treatment t is applied to z, and define $\theta_{tz} = \mu_{tz} - \mu_z$. Assume for simplicity that the y_z's are independent with a variance σ^2. Let η_z denote other parameters connected to the unit z.

In this situation it is natural to call $\phi = (\{\mu_z, \eta_z; z \in Z\}, \{\theta_{tz}; t \in T, z \in Z\}, \sigma^2)$ a c-variable for the system and $\Phi = \{\phi\}$ the c-variable space. This terminology is consistent with the one I introduced in Chapter 1 and with the one I will use in my approach to quantum mechanics in the next two chapters. Note that ϕ of course is not estimable in any conceivable experiment; nevertheless it is a useful conceptual quantity.

Let G be a transformation group defined on Z. This will induce a group on Φ. In other cases, larger groups on Z may be of interest, but in the case of designed experiments it is permutation of the experimental units which is the important issue.

Now for the experiment itself select a finite subset Z_0 of Z. We will assume for simplicity that G is so large that the full permutation group G_0 on Z_0 is a subgroup of G.

We will also assume that Z_0 is selected from Z by some random mechanism with the property that $\theta_t = \mathrm{E}(\theta_{tz}|t)$, expectation over this selection mechanism, is independent of the selected z. Then we will have for a given selected unit $z \in Z_0$ that

$$\mathrm{E}(y_z|t) = \mu_z + \theta_t.$$

This is one way to express the well known unit/treatment additivity, which is considered by Bailey [13, 14] and others to be crucial for having a consistent approach to the design of experiments.

From this point on Bailey [13] introduces an eight-stage experimental design theory, and this theory is developed further in [14]. We will only mention very briefly a few main points of this theory, referring to these and related papers for details. Note that Bailey's forthcoming book [15] seems to give a relatively full account of the field of experimental design, including the many important practical aspects of this area.

Block structure is an important aspect of experimental design theory and practice: Similar units are taken together in one block to enhance efficiency. This topic has many important facets, like Latin squares, split plot blocking, incomplete blocks and so on. From a group theoretical point of view, the main point is that the block structure determines the group used for randomisation: For a selected experiment $\mathcal{E}^a$, use for randomisation the largest subgroup G^a of G_0 which respects the block structure of that experiment: If the units z_1 and z_2 are in the same block, then $z_1 g$ and $z_2 g$ should be in the same block for all $g \in G^a$. The unit (names) are then randomised according to this group. This randomisation also has connections to the allocation of treatments.

Assuming that G^a is transitive, Bailey [14] proves the following: After randomisation, y_z (overusing this symbol slightly) has an expectation which only depends upon the treatment $t(z)$ given to z, and a covariance matrix C satisfying

$$C(z_1, z_2) = C(z_1 g, z_2 g), \qquad (3.21)$$

for $z_1, z_2 \in Z_0$ and $g \in G^a$. Using this, Bailey [14] introduces the *strata*, which are the eigenspaces of C, and which also are invariant spaces under the group G. The important practical point is that these give the lines of the (null) analysis of variance for the experiment, both in simple and in complicated cases.

3.11   Group Actions Defined on a c-Variable Space

Summary: A c-variable has been defined in Chapter 1 as a conceptual quantity connected to a system or a population. It can be illustrated by the set of parameters common to several experiments from which only one can be chosen. Typically a c-variable is inaccessible, that is, can not be estimated from any experiment. In this Section I let a group act on a c-variable space and look at subgroups associated with the single experiments. The issue is very relevant for the general approach to quantum mechanics in Chapter 5.

One advantage of the group approach is that it can be carried further than to a single experiment. Recall the Examples 1.10.4 and 1.10.5, where it was shown that it could be meaningful to define group actions on a c-variable space, even in cases where it only is possible to perform one of the potential experiments.

A similar conclusion could be drawn from the discussion in Section 3.8 above. And this conclusion will become crucial when we come to Chapters 4 and 5.

Let the total parameter space be Φ, and let a group G be acting upon Φ. For a given a, let $\mathcal{E}^a$ be an experiment that can be performed, and let Θ^a be the parameter space of that experiment. Then $\theta^a = \theta^a(\phi)$ is some given parameter value of the experiment $\mathcal{E}^a$, and everything that has been said earlier in this book is relevant to such an experiment.

In particular, then, to apply the theory of the present chapter, it is of interest to define groups acting upon each of the parameter spaces Θ^a. If this shall be induced from G on Φ, we have seen in Section 3.4 that it is necessary that the function $\theta^a(\phi)$ is permissible. But as pointed out in Lemma 3.4.1, this is always possible to achieve by taking a suitable subgroup.

Definition 3.11.1. *Let G^a be the maximal subgroup of G such that the function $\theta^a(\cdot)$ from Φ to Θ^a is permissible.*

Next we want to define group actions upon the observations corresponding to experiment $\mathcal{E}^a$, that is, on the sample space S^a. It is crucial then that (3.1) holds for each experiment. Note that we now go in the opposite direction than what we did in Section 3.2, where we started with the group actions on the sample space and then defined the group actions on the parameter space.

When starting with the group on the parameter space, the solution with respect to a sample space group is usually not unique. but there is one case where we get a unique solution. This is the situation where the experiment posesses complete sufficient statistics; recall definition (2.58).

Proposition 3.11.2. *Assume that there is a model Q^{θ^a} for experiment $\mathcal{E}^a$, that these measures are absolutely continuous with respect to each other, and that there for each a exists a complete sufficient statistic in the sample space S^a for this experiment. Let G^a act upon the parameter space Θ^a. Then there exist unique actions upon the space of complete sufficient statistics such that*

$$Q^{\theta^a g}(A) = Q^{\theta^a}(Ag^{-1}), \; g \in G^a. \tag{3.22}$$

The proof of Proposition 3.11.2 can be found in Appendix A.2.3.

3.12   Some Concluding Remarks

It might be appropriate here to cite from Efron [75]: 'A widely acceptable objective Bayes theory, which fiducial inference was intended to be, would be of immense theoretical and practical importance.'

One purpose of this chapter has been to clarify some problems connected to situations for which a fairly acceptable objective Bayes theory - a theory of optimal inference under invariance - is available. The price paid for this coherent theory seems to be two-fold: One has to fix a symmetry group for the problem at hand, and inference should preferably be limited to parameters that are permissible under this group.

As a conclusion to this chapter, adding a group structure to the model specification is of interest, and does have consequences. On the other hand, the symmetry approach to statistical inference also implies difficulties, most notably the difficulty of choosing a group in a given case. In general, the symmetries expressed by the group should have some substantial basis in the concrete problem described.

In many simple cases the choice of group is rather obvious, but it seems to be a challenge to find good, general rules for choosing the group in more complicated cases. Expressing lack of information in symmetry terms might be one way to proceed. Sometimes several groups lead to the same solutions.

Another question is whether the class of allowable parametric functions can be extended in any useful general way beyond the permissible ones. As illustrated in several cases above, however, this class can often be made rich enough for practical purposes by a suitable choice of group. As a general point, it must be more important to avoid incoherences than to be able to make inference on every possible parametric function.

Of course, then, at last: There are situations where it is not natural to choose a symmetry group at all before doing statistical inference, and there are other cases where it does not help much to choose a group at all even if this choice is made in a reasonable way. (Two such examples - related, and both attributed to C. Stein - can be found in [144] p. 231 and [24], p. 420.) A final open question is therefore whether any of the ideas in this chapter can be generalised also to certain specific situations in which it is difficult or useless to associate such a strong structure as a symmetry group.

Chapter 4

THE TRANSITION FROM STATISTICS TO QUANTUM THEORY

4.1 Theoretical Statistics and Applied Statistics

There is a large scientific literature within theoretical statistics. Many statistical journals also encourage good empirical papers, but by comparison the submission of purely applied papers to statistical journals seem to be relatively infrequent, and the publication in statistical journals of such papers without some methodological emphasis is even more infrequent. Yet, there is a huge activity related to applied statistics in a large number of fields in empirical science, either carried out by statisticians or by scientists educated in statistics.

There may be several explanations of the phenomenon above. To some extent one might say that there is a cultural gap between theoretical statisticians and applied statisticians. Some claim that the reason for this is that theoreticians use too much mathematics, and that applied people to a larger extent rely on intuition as a guideline.

This may be part of the explanation. On the other hand, to use mathematics as a precise language is undoubtedly very valuable in theoretical research, and also in the development of statistical methods which are useful for many applied areas. This has indeed proved to be the case on many occasions in the past.

To some extent one might say that theory and practice represent complementary values, but this should not prevent us from working towards better communication between research workers with different emphasis. Impulses from empirical investigations and problems encountered there may even lead to a broader mathematical basis for statistical theory.

In Chapter 2 I repeatedly said that statisticians sometimes ought to go beyond the standard statistical model, that is, a model where an experiment is thought to be synonymous with a class of probability measures. Fortunately, many theoretical investigations go beyond this frame; in modern data analysis this has been done several times with great success. But introductory courses at universities are almost exclusively confined to the standard statistical model. If one wants to explore the link towards other disciplines, a departure from the standard model will definitively be necessary.

Example 4.1.1. A normal confidence interval (Example 2.4.2) is a simple task to calculate, as are many related procedures in applied statistics (say one- and two-sample t-tests), and these procedures are used repeatedly in many different connections. In practice one must answer many difficult questions, however, in order to have a safe application of these procedures.

Formally, there are assumptions behind the procedures: Normality, independence and constant variance. There are methods to test these assumptions from data, but the results of such tests are always uncertain. Furthermore, the methods are to a certain extent robust against deviations from the assumptions: For instance, the pivot (2.19) will be approximately normally distributed by the Central Limit Theorem, even if the individual observations are not normal; even dependence or non-identical distributions may to a certain extent be tolerated. However, a single large deviation may ruin the argument.

Thus there are difficult questions to answer even in the simplest applications of statistics. In practice one can not avoid that some of these questions have to be approached intuitively. The final answer may depend on how large deviations from the formal assumptions one is willing to tolerate, and this again may depend upon the error one is willing to make in the test itself. Thus one ends up with a focused question and answer even in the simplest statistical test.

Example 4.1.2. In applied statistical investigations it is seldom enough to do just one experiment. Good experimental design books like Box, Hunter and Hunter [34] emphasize that we may learn in a stepwise way, often using the information from one experiment to design the next one. Nevertheless, in each experiment there are qualitative questions which different experimentalists answer differently ([34], p. 490)
- Different factors could have been chosen for the study.
- Different ranges for the factors could have been selected.
- Different choices could have been made for blocking factors.
- Different transformations for the factor might have been employed.
- Different responses and their metrics.
- Different models could have been considered.

The fact that each experimenter learns through several experiments, may serve to neutralize the effect of such subjective choices. In any case, many statistical investigations are carried out as a sequence of of focused questions. Several examples may be found in [34].

In my opinion it would be useful if our students could be exposed to the fact that theory is not pure mathematics; it must have an empirical basis. In an attempt to create a part of such a basis I included a large number of examples in Chapter 2. And, even though this may imply more use of mathematics, it was also meant as an extension of the basis for statistical theory to introduce the group aspect in Chapter 3. Here I even proposed to initially define a group on the c-variable space, a concept which definitely goes beyond the standard statistical model. This point of departure will be taken up again in Section 4.6.

4.2 The Gödel Theorem Analogy

One conclusion from our discussion of statistics so far is that the standard statistical model, while useful, to a certain extent must be considered to be incomplete from a practical point of view.

Some may feel that I draw an analogy too far in what I am going to say now, but I still feel that a good point can be made by mentioning this: In mathematical logic, a similar conclusion was reached long time ago. A well known general theorem by Kurt Gödel from 1931 – see [101] – says that every rich enough theory may be regarded as incomplete in a certain sense.

More precisely, Gödel demonstrated that within any rich enough branch of mathematics, there would always be some propositions that could not be proven either true or false using the rules and axioms of that mathematical branch itself. You will not be able to prove every conceivable statement about numbers within a system without going outside the system in order to come up with new rules and axioms. And then by doing so you will only create a larger system with its own unprovable statements. The implication is that all logical system of any complexity are, by definition, incomplete; each of them contains, at any given time, more true statements than it can possibly prove according to its own defining set of rules.

True, this is a result about pure mathematics, but at least by analogy our tentative statement is that we can say something similar about systems involving empirical sciences, too: Any formal set of mathematical axioms will be incomplete in the sense that problems occur in practice that can not be discussed properly strictly within the frames of these axioms. If this can be taken as a general statement, it could have far-reaching implications for the way theory should be carried out.

In my view, there might be said to be an analogous situation for quantum theory. There is a welldefined set of axioms for the theory, which we will formulate shortly, and which most physicists agree upon. Yet, there is still an intense debate about the foundation of the theory, something that can be most easily seen by sampling a few of the many monthly papers on the web system quant-ph at http://xxx.lanl.gov/.

In fact, Albert Einstein was all of his life sceptical towards the whole of quantum mechanics, and his most pressing argument was that the theory must be incomplete, see Einstein, Podolsky and Rosen [74]. I will come back to this issue in Chapter 6.

What I will do in this book, is in essence to formulate an axiomatic basis which covers important parts of both statistical theory and quantum theory. In agreement with the discussion above, one should not expect this basis to be complete, either, even though it may explain more phenomena than the old basis. One will always encounter cases where intuition is needed in addition to formal rules.

4.3   Wave Mechanics

Some of the basic concepts of quantum mechanics are wave functions, operators and eigenvalues of the operators. Following Isham [130], I will start by introducing these concepts in wave mechanics, where the operators can be looked upon as differential operators. Later we will look at electron spins and related systems, where the operators are complex matrices. Most operators, both in wave mechanics and in discrete systems are self-adjoint, which means that their eigenvalues are real-valued.

Definition 4.3.1.

a) A number a is an eigenvalue of a differential operator T if it satisfies the differential equation

$$Tu(x) = au(x)$$

plus appropriate boundary conditions on the function $u(x)$, for instance that it is square integrable: $\int_{-\infty}^{\infty} |u(x)|^2 dx < \infty$.

b) A self-adjoint operator T is one for which

$$\int_{-\infty}^{\infty} (T\psi)^*(x)\phi(x)dx = \int_{-\infty}^{\infty} \psi^*(x)(T\phi)(x)dx$$

for all square integrable wave functions ψ and ϕ, where the symbol $*$ denotes complex conjugations. Or more precisely: The adjoint $T^\dagger$ of an operator T is defined by

$$\int_{-\infty}^{\infty} \psi^*(x)(T^\dagger\phi)(x)dx = \int_{-\infty}^{\infty} (T\psi)^*(x)\phi(x)dx,$$

and self-adjointness is simply $T = T^\dagger$.

On this background we can formulate 4 quantum rules for wave mechanics, rules that will be formalised in the next section:

Rule 1. The quantum state of a point particle moving in one dimension is represented by a complex-valued wave function $\psi(x)$ that can be normalised to one:

$$\int_{-\infty}^{\infty} |\psi(x)|^2 dx = 1.$$

A crucial idea in quantum theory is that for virtually any pair of wave functions ψ_1 and ψ_2 one can take complex normalised linear combination and form new wave functions.

Rule 2. Any physical quantity A that can be measured is represented by a self-adjoint differential operator T that act on the wave functions.

Rule 3. The only possible results of measuring A is one of the eigenvalues of the operator T which represents it.

Assume for simplicity that T has a discrete set of non-degenerate eigenvalues $a_1, a_2, \ldots$. Let the quantum state be $\psi(x)$. Then the probability that A takes the value a_n is given by

$$P_\psi(A = a_n)) = |\psi_n|^2,$$

where the complex numbers ψ_n are the coefficients in the expansion

$$\psi(x) = \sum_{n=1}^\infty \psi_n u_n(x)$$

of the wave function $\psi(x)$ as a linear combination of the normalised eigenfunctions of T.

As we see, even in the rather concrete setting of wave mechanics, we have to except some rather formal rules in other to develop an operational quantum theory. The good side of this is that these rules function in a large number of settings, and give results in agreement with experiments. Wave mechanics was first developed by Schrödinger, and was later shown to be equivalent with Heisenberg's matrix machanics.

There is a fourth rule, which I will come back to later, which gives the time-development of the wave function $\psi(x)$: The Schrödinger equation.

Let us go back to some consequences of Rule 3. It can be shown as a conseqence of self-adjointness and non-degeneracy that the eigenfunctions $u_n(x)$ form a complete set, that is, any wave function $\psi(x)$ can be expanded uniquely as above. Furthermore, the normalised eigenfunctions satisfy the orthonormality condition

$$\int_{-\infty}^\infty u_m^*(x) u_n(x) dx = \delta_{mn},$$

which implies

$$\psi_n = \int_{-\infty}^\infty u_n^*(x) \psi(x).$$

Finally, orthonormality implies:

$$\int_{-\infty}^\infty \psi^*(x) \phi(x) dx = \sum_{n=1}^\infty \psi_n^* \phi_n,$$

in particular

$$\int_{-\infty}^\infty |\psi(x)|^2 dx = \sum_{n=1}^\infty |\psi_n|^2,$$

and the following obvious relation follows:

$$\sum_{n=1}^\infty P_\psi(A = a_n) = 1.$$

Thus the quantum formulation leads to a proper probability distribution. The expectation in this distribution is

$$E_\psi(A) = \int_{-\infty}^{\infty} \psi^*(x)(T\psi)(x)dx.$$

To complete the calculating rules for wave mechanics, it is left to specify how the operator T should be constructed from the observable A. Since this implies some non-trival mathematical problems, and since these problems are discussed in [163] and other mathematically oriented books on quantum mechanics, I abstain from going into this question in detail here. I only mention that the position observation x is always translated into the multiplication operator x, while the momentum observator p is translated into the differential operator $\frac{\hbar}{i}\frac{d}{dx}$. Already this leads to convergence problems, which are neglected in much of the the physics literature. Even more severe problems may occur when one wants to construct an operator for $A(x, p)$ by substitution, since the two operators involved do not commute. In the rest of this book we will assume that the operator T for some observable A is given.

Wave mechanics can be looked upon as one of the simplest realisations of quantum mechanics. The states are represented by wave functions $\psi(x)$ satisfying $\int_{-\infty}^{\infty} |\psi(x)|^2 dx = 1$. This is the Hilbert space of wave mechancs. In general the states of quantum mechanics are given by vectors v of some Hilbert space $\mathbf{H}$, a complete vector space with a scalar product. The vectors v are adjoined by vectors $v^\dagger$, corresponding to $\psi^*(x)$, and they are normed by $v^\dagger \cdot v = 1$.

4.4   The Formal Axioms of Quantum Theory

The ordinary axiomatic basis of quantum mechanics can be formulated in a quite straightforward mathematical way. Thus on this basis there is no reason to criticise the established theory. Also, the theory seems to be valid in the sense that it has been confirmed by numerous experiments.

My criticism will be concentrated on two points: 1) The ordinary axioms are purely formal. 2) There is nowhere in these axioms a hint about any connection to statistical theory, the other major probability based theory for prediction of new observations from some state determined by previous observations.

Specifically, I will consider the following axioms of quantum theory, taken from Isham [130]. Then I will indicate a possible new basis more closely related to the statistical theory discussed in this book, and in Chapter 5 and Chapter 6 I will among other things derive Isham's rules from this setting.

Quantum rule 1. *The predictions of results of measurements made on an otherwise isolated system are probabilistic in nature. In situations where the maximum amount of information is available, this probabilistic information is represented mathematically by a vector in a complex Hilbert space $\mathbf{H}$ that forms the state space of the quantum theory. In so far as it gives the most precise predictions that* *are possible, this vector is to be thought of as the mathematical representation of the physical notion of 'state' of the system.*

Quantum rule 2. *The observables of the system are represented mathematically by self-adjoint operators that act on the Hilbert space* **H**.

Quantum rule 3. *If an observable quantity $\hat{\lambda}^a$ is represented by the self-adjoint operator T^a, and the state by the normalised vector $v \in$* **H***, then the expected result of the measurement is*

$$E_v(\hat{\lambda}^a) = v^\dagger T^a v. \tag{4.1}$$

Here $v^\dagger$ is the transpose and complex conjugate of the vector v.

Quantum rule 4. *In the absence of any external influence (i.e., in a closed system), the state v_t changes smoothly in time t according to the time-dependent Schrödinger equation*

$$i\hbar \frac{dv_t}{dt} = H v_t, \tag{4.2}$$

where $i = \sqrt{-1}$, $\hbar$ is Planck's constant and H is a special operator known as the Hamiltonian.

One way to formulate the essence of the first three rules, is that the state of an isolated system is given by a vector v in some Hilbert space, and that probability statements can be found from calculations using this vector.

Now a vector v can always – in fact in many ways – be regarded as an eigenvector of some operator T. Assume that one such operator is physically meaningful and corresponds to some physical quantity λ. Then there will be an eigenvalue λ_k corresponding to the eigenvector v of T, and from Rule 3 one can easily show that the state v is characterised by the statement that any measurement of the quantity λ gives the result λ_k with certainty. (Later we will call the measurements of Rule 3 perfect measurements, since they do not take into account measurement errors.)

Namely, from (4.1), the expectation of the measurement is λ_k, and the variance is

$$v^\dagger (T - \lambda_k)^2 v = 0. \tag{4.3}$$

On the other hand, if a state vector v is characterised by the fact that any measurement of λ gives a certain value λ_k, then it follows from (4.3) that v must be an eigenvector of T with eigenvalue λ_k.

In general, the physical quantity λ can be composed of several quantities like the charge and a spin component for one particle, or such quantities for each particle if several particles are involved. If this list of quantities is large enough, the vector v will be a unique eigenvector (except for an irrelevant phase factor) of the corresponding compound operator T.

In this way, the state v is characterised by the fact that there is a certain physical quantity that has a certain value in the state v. (This quantity may consist of several parts; in this respect it is assumed to be maximal.) In other words: The state is characterised by two components: 1) A maximal question (which can consist of several parts): What is the value of λ^a? (For a precise definition of maximality, see Section 4.7 below.), and 2) The answer to this question, here given by the eigenvalue (incorporating the values of the physical quantities of the parts).

In this chapter and the next one, we will show that this characterisation in a large number of cases can be inverted: Start with certain reasonable assumptions, some of which are motivated from the previous chapters in this book. Then say that a state of some system is given by 1) A question, 2) The certain answer of that question. Then I will show that the Hilbert space above may be constructed, and the question-and-answer combinations stand in a one-to-one correspondance with the vectors of that Hilbert space. Thus it seems that it is possible to construct the formal world of quantum theory from what, at least conceptually, is a relatively non-formal idea.

The assumptions that are needed for this, are given in Sections 4.7 and 4.8 below. The most important ones are the following:

- There exists a c-variable space for the system in question, and a transitive transformation group defined on this space.

- For any experiment, the context is so limited that only certain functions of the c-variable can be given experimental values.

- Different such estimable parametric functions may be connected together through elements of the group.

These assumptions will be more precisely formulated as axioms in Section 4.7 below and used in detail to develop the theory in Chapter 5.

4.5   The Historical Development of Formal Quantum Mechanics

The earliest book on the mathematical foundation of quantum mechanics is von Neumann [162]; in English translation: von Neumann [163]. This book has had a great influence; in its time it constituted a very important mathematical synthesis of the theory of quantum phenomena. The book can also be considered to be a forerunner of quantum probability. For physicists, von Neumanns book was supplemented by the book of Dirac [63], which started the development leading to modern quantum field theory.

The development of quantum probability as a mathematical discipline, continuing the more formal development of quantum theory, was started in the 1970s. A first important topic was to develop a noncommutative analogue of the notion of stochastic processes; see Accardi [4] and references there. Other topics were noncommutative conditional expectations and quantum filtering and prediction theory (Belavkin [21] and references there).

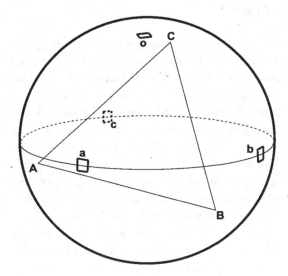

Fig. 4.1 The large scale model.

Quantum probability was made popular among ordinary probabilists by Meyer [157]. A related book is Parthasarathy [169], which discusses the quantum stochastic calculus founded by Hudson and Parthasarathy, but also many other themes related to the mathematics of current quantum theory. An example of a symposium proceeding aiming at covering both conventional probability theory and quantum probability is Accardi and Heyde [5].

There are also links between quantum theory and statistical inference theory. A systematic treatment of quantum hypothesis testing and quantum estimation theory was first given by Helstrom [120]. In Holevo [125] several aspects of quantum inference are discussed in depth; among other things the book contains a chapter on symmetry groups. A survey paper on quantum inference is Malley and Hornstein [149].

As an example of a particular statistical topic of interest, consider that of Fisher information. Since a quantum state ordinarily allows several experiments, this concept can be generalised in a natural way. A quantum information measure due to Helstrom can be shown to give the maximal Fisher information over all possible experiments; for a recent discussion see Barndorff-Nielsen and Gill [17].

One can thus point at several links between ordinary probability and statistics on the one hand and their quantum counterparts on the other hand. However, a general theory encompassing both sides, based on a reasonably intuitive foundation, has so far been lacking.

4.6   A Large Scale Model

In the next section I will start on formulating my axioms for quantum mechanics. As a preparation for this, and partly as a motivation, I will first give a largescale example, a system where much of these basic axioms can be seen to be satisfied.

One of the simplest non-commutative groups is the group S_3 of permutations of 3 objects. It has a two-dimensional representation discussed in many books in group theory and in several books in quantum theory. The quantum theory book by Wolbarst [208] is largely based upon this group as a pedagogical example.

In this section I will visualise this group by considering the permutations of the corners of an equilateral triangle, which can be realised physically by the change of position of some solid version of this triangle. This will serve to illustrate the quantum foundation below on a macroscopic example.

The spatial orientation ϕ of the whole triangle will be looked upon as a hidden c-variable, and to this end we will imagine the solid triangle placed within a hollow non-transparent sphere, with the corners on the sphere, in such a way that it can rotate freely around its center point, placed at the center of the sphere. The basic group G is to begin with taken as the group of such rotations, but later, when we specialise to the corners, we will take G as the permutation group. Let the solid triangle be painted white on one side and black on the other side.

Let there be 4 small windows in the sphere, one at the north pole, where the colour facing up can be observed, and three equidistant windows along the equator, where the closest corner of the triangle can be observed.

Every observation is made through one and only one of these windows, and thus we are never able to obtain a complete picture of what is inside. In this way the triangle as described above must have the status of a mental image, and the position of the triangle must be regarded as inaccessible.

To make a connection to ordinary statistical inference, the measurements made in the windows could be uncertain for some reason, and we could model this in the ordinary statistical way by some model $Q^\lambda(dy)$, depending upon a (reduced) parameter which can be thought about as the ideal measurement in one particular window.

Hence there are 4 reduced parameters, corresponding to the 4 different experiments that can be done in this case, one for each window: λ^0 is the ideal colour as observed from the north pole window: λ^a, λ^b and λ^c are the three 'correct' corners of the triangle as observed from the windows a, b and c, respectively. The term 'correct' will be defined more precisely below. The parameter λ^0 takes the values *black* and *white*, and the parameters λ^i for $i = a, b, c$ each takes the values A, B and C, say. All these parameters can be considered to be functions of the triangle's spatial orientation ϕ within the sphere.

Lemma 4.6.1. *Both with respect to the group of permutations and with respect to the group of rotations, λ^0 is a permissible function, while λ^a, λ^b and λ^c each* *are non-permissible. The largest permutation group with respect to which λ^a is permissible, is the group G^a of cyclic permutations of the corners of the triangle, similarly for λ^b and λ^c.*

Proof.

Consider the 6 elements of the group S_3 of permutations: $g_1(ABC \mapsto ABC)$, $g_2(ABC \mapsto CAB)$, $g_3(ABC \mapsto BCA)$, $g_4(ABC \mapsto ACB)$, $g_5(ABC \mapsto CBA)$, $g_6(ABC \mapsto BAC)$.

Assume $\lambda^0(\phi_1) = \lambda^0(\phi_2)$, say black, for two c-variable values ϕ_1 and ϕ_2. Then by simple inspection, $\lambda^0(\phi_1 g_i) = \lambda^0(\phi_2 g_i) = black$ for $i = 1, 2, 3$ and *white* for $i = 4, 5, 6$. Hence λ^0 is permissible.

For the other functions it is enough to produce a counterexample. Here is one for λ^a: Let ϕ_1 be any hyperparameter value, and by definition let ABC be the sequence of corners in ϕ_1 corresponding to the windows a, b, c. Put $\phi_i = \phi_1 g_i$ for $i = 2, \ldots, 6$. Then $\lambda^a(\phi_1) = \lambda^a(\phi_4) = A$, but $\lambda^a(\phi_1 g_5) = \lambda^a(\phi_5) = C$ and $\lambda^a(\phi_4 g_5) = \lambda^a(\phi_3) = B$. The group elements g_4 and g_6 have the same structure as g_5: permutation fixing one corner of the triangle. Therefore a similar statement holds for these group elements.

To check that λ^a is permissible under the cyclic group, we can use direct verification. The details are omitted. A geometric proof is simpler than an algebraic proof.

One can easily imagine that an ideal measurement at the window a in principle can give more information than λ^a about the position of the triangle, but this information is hidden. One way to make this precise, is the following: Let us divide the sphere into 3 sectors corresponding to each of the windows a, b and c by using the meridians midway between a and b, midway between a and c and midway between b and c as borders between the sectors. Let S^a be the sector containing window a. Define θ^a as 1) the points among the triangle corners A, B and C that happen to be in the sector S^a; 2) the coordinates of any two points which happen to belong to the same sector.

From the geometry, it can be seen that S^a can contain 0, 1 or 2 triangle corners. This can be used to define λ^a precisely: If S^a contains 1 corner, let this corner be λ^a. If S^a contains 2 corners, let the closest one, as calculated from the coordinates, be λ^a. If S^a contains 0 corners, then exactly one of its neighboring sectors must contain 2 corners. One is then chosen to be λ^b, respectively λ^c; let the other one, the closest one to the window a, be λ^a. Since the coordinates of the corners that are in the same sector are contained in θ^a, it is seen that λ^a is a function of θ^a.

Note that the reduction from ϕ via θ to the parameter λ is forced upon us in this situation by the limitation in the possibility to make observations on the system.

We assume that there is some mechanism to ensure that it is impossible to look through two equatorial windows at the same time.

4.7   A General Definition; A c-System

Motivated by the previous section and by the preceding discussion it is natural now to describe a fairly general closed system and to formulate certain properties which such a system ought to have in order to resemble both the system in the example above and certain simple quantum systems. For simplicity, I will call such a system a conceptually defined system or a c-system.

Definition 4.7.1.   *A c-system is a closed system for which there exists a c-variable space Φ whose elements $\phi \in \Phi$ are not estimable relative to any experiment, that is, they are inaccessible. There is a transformation group G defined on Φ. The space Φ is locally compact, and the group G is transitive on Φ. The right invariant measure under G is called ν.*

Furthermore, there exists a set $\mathcal{A}$ of potential experiments $\mathbf{E}^a; a \in \mathcal{A}$ on this system. For each $a \in \mathcal{A}$ there is a maximal estimable parameter λ^a. More precisely, maximal means maximal relative to the partial ordering: $\lambda << \theta$ iff there is a function h such that $\lambda = h(\theta)$.

When λ^a is discrete, there exists a perfect experiment for λ^a where measurement noise can be disregarded and where the resulting (pure) states are given by statements of the form $\lambda^a = \lambda_k^a$.

The definition of maximality is consistent with the fact that each c-variable ϕ is inaccessible, while each ordinary parameter is accessible. The definition itself is also consistent: When θ is accessible, then so is $\lambda = h(\theta)$. And if λ should be inaccessible with $\lambda = h(\theta)$, then θ is also inaccessible.

Those not interested in technical mathematics should just regard Φ as some general space. The requirement that it is locally compact, and hence in particular a topological space, is just what is needed to obtain an invariant measure. For simplicity one can for instance think of Φ as a sphere, and ν as the constant measure on that sphere. But very many possibilities exist.

A perfect experiment means that the estimator of the parameter is exactly equal to the parameter itself; thus the statistical inference is trivial. This limiting case will correspond to the simplest case of the quantum-mechanical formalism, but cases that are more interesting from a statistical point of view involve a non-trivial measurement apparatus. In the perfect experiment case the value obtained by the parameter is exactly what is observed.

Notice that a statement of this last form defines a state for the system. When I later discuss transition probabilities, the defining state will be of this form. The prerequisite for this is that λ^a is maximal in a the sense that the transition can be regarded as a Markov process in these variables. Note that the state consists of two elements: A question $a \in \mathcal{A}$: What is λ^a? - together with an answer λ_k^a. In the triangle-in-a-sphere example of section 4.5 the question consists of looking through a given equatorial window. In the quantum-mechanical case I will also come back to this characterisation.

Lemma 4.7.1. *The triangle-in-a-sphere system of Section 4.6 is a c-system.*

Proof.
Obvious.

As in this particular system, it is often natural to introduce intermediate parameters $\theta^a = \theta^a(\phi)$ which may not be estimable under the experiment $\mathbf{E}^a$. Then $\lambda^a = \lambda^a(\theta^a)$ is estimable.

4.8 Quantum Theory Axioms under Symmetry and Complementarity

I am now ready to formulate an alternative set of axioms of quantum theory, axioms that may be motivated from the statistical discussion above, and will cover many quantum systems. I will aim at deriving the Quantum rules 1-3 of Section 4.4 from these axioms in the next Chapter, and in fact, I will derive more than that.

In contrast to the traditional set given by Quantum rules 1-3, I consider most of the axioms below to be relatively natural in the light of common sense and in the light of basic statistics in the way it has been formulated above. It is the totality of assumptions which makes the core of quantum theory in the way I see it.

For the rest of this section I assume some given c-system. I am not interested in considering the most general quantum theory here; I will limit everything to systems with symmetry as given by the group G. This will include particles with arbitrary spin and systems of such particles, and is enough to illustrate the connection between statistics and quantum theory. In my view, this connection goes in both directions, as it gives a possibility for both disciplines to learn from the other. In the following, I will formulate some axioms that a quantum-mechanical c-system must satisfy.

In the definition of the c-system, a perfect experiment was mentioned. For the case when the experiment is not perfect, it is natural to assume Bayesian estimation in each experiment with a prior induced on each λ^a by ν on Φ. This is also equivalent to what is called the best equivariant estimator under the group G^a defined below, see Chapter 3. The existence of this group is ensured by Lemma 3.4.1. The uniqueness of the best equivariant estimator is ensured, since the invariant measure ν on Φ is unique by the transitivity of G on Φ.

Definition 4.8.1. *Let G^a be the maximal subgroup of G with respect to which $\lambda^a(\phi)$ is a permissible function of ϕ.*

In the next axiom we assume that the experiments corresponding to different parameters λ^a can be connected by group transformations.

Axiom 1. *For each pair of experiments* $\mathbf{E}^a, \mathbf{E}^b; a, b \in \mathcal{A}$ *there is an element* g_{ab} *of the basic group* G *which induces a correspondence between the respective reduced parameters:*

$$\lambda^b = \lambda^a g_{ab} \text{ or } \lambda^b(\phi) = \lambda^a(\phi g_{ab}). \tag{4.4}$$

From Axiom 1 we will prove in the next Chapter that $g^a = g_{ab}g^b g_{ba}$. What we really need there is the implied relation between unitary group representations (for details see definition in Chapter 5)

$$U(g^a) = U(g_{ab})U(g^b)U^{-1}(g_{ab}). \tag{4.5}$$

Briefly, a group representation is a set of operators $U(g)$ such that $U(gh) = U(g)U(h)$. In this connection we can take $U(g)$ as the right regular representation defined on $L^2(\Phi, \nu)$ by:

$$U(g)f(\phi) = f(\phi g).$$

The symmetry assumptions between experiments given in Axiom 1 will turn out later to be crucial for the development of quantum theory. The main purpose of this axiom is to enable us to construct in a natural way the Hilbert space of quantum mechanics. The states, i.e. the amount of knowledge of a given system obtained through a perfect experiment by asking a maximal focused question: What is λ^a? together with the answer: $\lambda^a = \lambda^a_k$, are represented by unit vectors of that Hilbert space.

More specifically: We will show in Chapter 5 that these axioms lead to the existence of a Hilbert space $\mathbf{H}^a$ for the experiment $\mathbf{E}^a$ in such a way that there are basis vectors $f^a_k \in \mathbf{H}^a$ which can be uniquely coupled to the statements that a perfect measurement of λ^a gives the result that $\lambda^a = \lambda^a_k$ for suitable constants λ^a_k. These Hilbert spaces can be realised as subspaces of $L^2 = L^2(\Phi, \nu)$, where ν is the invariant measure under G assumed in Axiom 1. In particular, $\mathbf{H}^a$ is given by

$$\mathbf{H}^a = \{f \in L^2 : f(\phi) = \tilde{f}(\lambda^a(\phi))\}, \tag{4.6}$$

where $\tilde{f}$ is some function. Using (4.5), we will define a common Hilbert space $\mathbf{H}$ such that all the spaces $\mathbf{H}^a$ are unitarily related to $\mathbf{H}$. The following axiom is needed for further properties of this basic space.

Axiom 2. *The group* G *is the smallest group containing all the subgroups* $G^a; a \in \mathcal{A}$.

So far, I have not said anything about the nature of the parameters λ^a. For the most part in this book I will confine myself to standard quantum mechanics where the observables have a discrete spectrum. In our setting this corresponds to the following assumption:

Fig. 4.2 Thought model of an electron.

Axiom 3. *Each parameter λ^a assumes a finite or denumerable set of values.*

Finally, we introduce quantum probabilities. This is done by proving a cele-brated formula of Max Born by taking as a point of departure the following axiom, essentially a symmetry property of the transition probabilities from one perfect experiment to another:

$$P(\lambda^b = \lambda_i | \lambda^a = \lambda_k).$$

Axiom 4. *(i) The transition probabilities are well-defined in the sense that the probabilities above do not depend upon anything else.*

(ii) The transition probability from $\lambda^a = \lambda_k$ in the first perfect experiment to $\lambda^a = \lambda_k$ in the second perfect experiment is 1.

(iii) For all a, b, c we have that $\mu(\phi) = \lambda^a(\phi g_{bc})$ is a valid experimental parameter.

(iv) For all a, b, c, i, k we have

$$P(\lambda^b(\phi) = \lambda_i | \lambda^a(\phi) = \lambda_k) = P(\lambda^b(\phi g_{bc}) = \lambda_i | \lambda^a(\phi g_{bc}) = \lambda_k)$$

In Chapter 5 all these axioms will be repeated and commented upon as they are used to develop the link to quantum theory.

Proposition 4.8.1. *The triangle-in-a-sphere system of Section 4.6 satisfies the axioms 1, 3 and 4, but not Axiom 2.*

<u>Proof.</u>
The group elements g_{ab} of Axiom 1 are formed by rotation of the triangle. Axiom 3 is obvious. The Markov property of (i) of Axiom 4 has already been considered and is obvious here; the transition probabilities of (iv) are $1/2$ here. The group G^a consists of the identity plus the interchange of b and c. It is impossible to form rotations by combining such interchange group elements, so Axiom 2 does not hold.

4.9   The Electron Spin Example

The most simple quantum mechanical system, a qubit, is realised as an electron with its spin. The spin component λ can be measured in any spatial direction a, and λ always takes one of the values -1 and $+1$.

In this section, I will give a non-standard, but quite intuitive description of a particle with spin, a description which we later will show to be equivalent to the one given by ordinary quantum theory.

Look first at a general classical angular momentum. A c-variable ϕ corresponding to such an angular momentum may be defined as a vector in three dimensional space; the direction of the vector giving the spin axis, the norm gives the spinning speed. A possible associated group G is then the group of all rotations of this vector in $\mathcal{R}^3$ around the origin, possibly with scale changes included.

Now let the electron at the outset have such a c-variable ϕ attached to it, and let $\kappa = \|\phi\|$. It is well known that it is impossible to obtain in any way such detailed information about the electron spin, but this definition is consistent with our concept of an inaccessible c-variable. Continuing this hypothetical description, assume that we set forth in an experiment $\mathcal{E}^a$ to measure part of its angular momentum component $\theta^a(\phi) = \kappa \cos(\alpha)$ in some direction given by a unit vector a, where α is the angle between ϕ and a. The measurement can be thought of as being done with a Stern-Gerlach device, which strictly speaking measures an observable y whose distribution depends upon (some function of) θ^a, implying a possibility that the parameter θ^a - or some part of it - can be estimated from such a measurement. Given a, and given the measurement in the direction a, the rest of the c-variable ϕ will be unestimable.

With respect to the group G, the function $\theta^a(\cdot)$ is easily seen to be non-permissible for fixed a, simply because two vectors with the same component along a in general will have different such components after a rotation. The maximal possible group G^a with respect to which θ^a is permissible, is the group generated by the rotations of the vector ϕ around the axis a possibly together with a $180°$ rotation around any axis perpendicular to a, plus a possible scale change $\kappa \mapsto b\kappa; b > 0$.

In analogy to the situation in Section 4.6, assume now that the electron's c-variable ϕ always is hidden, in such a way that for every a, the only part of $\kappa \cos(\alpha)$ we are able to measure, is the value $+1$ or -1, giving the sign of this component. We call this part $\lambda^a(\phi)$. This is an extreme model reduction compared to θ^a, but interestingly enough, the model reduction is to an orbit of the group G^a.

The measured part found by some measurement apparatus, which may contain additional measurement noise, also takes the values ± 1, and is called $\hat{\lambda}^a$. In some instances below, we will disregard such measurement noise, and assume the ideal condition $\hat{\lambda}^a = \lambda^a$. Such an approximation makes sense, also from a statistical point of view, for a discrete parameter.

Finally, since the model reduction is to a parameter of fixed norm, we delete the scale change part from the groups G and G^a. In particular, G is the group of rotations.

Proposition 4.9.1. *The electron spin system is a c-system which satisfies axioms 1-4.*

Proof.
Obvious from the discussion above.

A general discussion on how to construct a c-variable from a series of mutually exclusive experiments, illustrated by the case of electron spin, is given in Appendix A3.1.

Chapter 5

QUANTUM MECHANICS FROM A STATISTICAL BASIS

5.1 Introduction

[1]Nancy Cartwright [40] has argued that physical laws are about our models of reality, not about reality itself. This is a statement with severe implications. At the same time she argues (p. 186) that the interpretation of quantum mechanics should be seen entirely in terms of transition probabilities. The present book is in agreement with both these statements. I will supplement the first statement, however, by saying a few words about models. For many reasonably complex phenomena, several models, and several ways to give a language for model formulation, can be imagined. In some instances these models, while appearing different, are so closely related that they give the same predictions about reality. If this is the case, I think many would agree with me that we should choose the model which has the most intuitive interpretation, also in cases where there may be given strong historical and culturally related arguments for other models. It may of course be the case that the conventional model is more suitable for calculations, but this should not preclude us from employing more intuitive points of view when arguing about the model and when trying to understand complex phenomena.

In the present chapter I take as a point of departure such a different, in my view more intuitive, modelling approach to quantum mechanics, first formulated as a definition in Section 4.6 supplemented by 4 axioms in Section 4.7. I will show that the ordinary quantum formalism, at least to a large extent, follows from this. I limit myself to the time-independent case here; the time-dependent case will be taken up in Chapter 6. The essence of the axioms is of statistical nature in the sense that they are related to parameters of potential experiments.

To start such a programme, we have to relate it to the usual formulation of quantum theory. The basic quantum theory formulation is given in different ways

---

[1]This chapter is somewhat mathematical. Those who feel uncomfortable with this mathematics, can look quickly through the results in the first reading, although some of these are essential to understand in order to appreciate the main development and the connections to parts of the rest of the book. As a very minimum, the introduction together with the summary of each section should be read. Needless to say, those wanting to understand the principles involved, should spend time also on the details.

in different books, although all agree about the foundation. For definiteness I will in this chapter consider the three rules taken from Isham [130] and reproduced in Section 4.3 as Quantum rule 1 - Quantum rule 3. In the last parts of this chapter I will give a derivation of them (and in fact more) under the assumptions given in Sections 4.6 and 4.7.

These three rules together with the Schrödinger equation (Quantum rule 4) constitute according to [130] the basic assumptions of quantum mechanics. The axioms which I use to derive these rules are in essence a natural extension of statistical theory. I will also comment upon issues like superselection rules and the extension to mixed states, and try to argue that formal equations like (4.1) can be associated with a very natural interpretation. A basic step in the derivation of the three rules is to give arguments for (a variant of) Born's celebrated formula for transition probabilities in quantum mechanics.

Note that the states of conventional quantum mechanics, in the non-degenerate case, can be interpreted in the following way: Corresponding to the operator T there is a physical variable λ, and a state, as given by a state vector v_k connected to a particular value λ_k of λ. The vector v_k is an eigenvector of T corresponding to the eigenvalue λ_k. I will assume here that the eigenvalue is nondegenerate, which corresponds to a variable λ which is maximal: It is not a non-trivial function of any other measurable variable connected to the system.

Given the variable λ, the operator T is determined from physical considerations, and given T and the eigenvalue λ_k, the eigenvector v_k is determined. Thus we can say that the state is determined by two elements: 1) A maximal question: What is the value of λ? 2) The answer: $\lambda = \lambda_k$. In this chapter I show that we can go the other way, and start with a question/ answer pair like above, and then under certain assumptions, mainly related to symmetry and to a limited context, we arrive at the Hilbert space formulation with the above interpretation of the state vectors.

Consider a particle with spin 1/2, like an electron, and let a be any 3-vector. Then it makes sense to formulate the question: What is the spin component in direction a? together with the answer: +1. And it makes sense to say that these two elements together define a state; in the common quantummechanical language the state is a vector in the two-dimensional complex vector space, a vector with eigenvalue +1 for the operator $a \cdot \sigma$, where a is the direction vector above, the ordinary scalar product is used, and where σ is the vector composed of the three Pauli spin matrices:

$$\sigma_x = \begin{pmatrix} 0 & 1 \\ 1 & 0 \end{pmatrix}, \ \sigma_y = \begin{pmatrix} 0 & -i \\ i & 0 \end{pmatrix}, \ \sigma_z = \begin{pmatrix} 1 & 0 \\ 0 & -1 \end{pmatrix}. \tag{5.1}$$

My aim is to approach this in a less formal way than the state vector approach, usually called the Hilbert space approach. To carry out such a programme, I want to use other, more direct, mental models.

The point is that I consider the spin vector in a model using what I earlier in this book have called the c-variable concept (see Sections 1.2, 1.3 and 4.6). An inaccessible c-variable is something that can be formulated by ordinary language and is associated with a tentative model of the subatomic reality, but for which it is far too optimistic to connect a definite value in the sense that this value can be estimated by an experiment. The spin vector will be thought of as such an inaccessible c-variable. A general approach towards c-variables taking mutually exclusive experiments as a point of departure, is given in Appendix A3.1.

What we can confront with experiments, though, is some given function of this spin vector, in the spin 1/2 case the sign of the component of the spin vector in a certain direction. The rest of the spin vector will always remain unknown to us. One might of course say that then it is nonsense to speak about the rest of the spin vector, but I would say that it is useful to have a mental model. Various people may have different mental models, but this does not matter as long as they agree about the symmetry aspect and about the observable part of the spin vector.

A relatively concrete realisation of this is given by the triangle-in-a- sphere which I have described in Section 4.5, for which we are only allowed to look through one chosen window. Here we may have constructed the triangle with a given colour, made it from a definite material and so on, but to an observer in a window, all these parameters may only be imagined mentally. All that matters for his observations are the corners A, B or C, and all that matters in order to interpret these observations is a mental model of a triangle with rotational symmetry attached to it.

Or consider the case of a single patient at a fixed time (Example 1.3.4), where we might be interested in expected recovery times τ_1 and τ_2 under two potential treatments (and in other parameters), so that the vector $\tau = (\tau_1, \tau_2)$ does not have an empirical value, but one component can be estimated by an experiment.

But I repeat: The c-variable may nevertheless be a useful quantity. In the model context it may help us to just have a mental picture of what we think is going on, say, in the subatomic world. The spin vector can be red or blue, can be imagined to be connected to some solid body, or just be an arrow. But what is the same in all these mental pictures, is every single component that we are allowed to ask questions about.

One very useful property of the c-variable is that, even when it is inaccessible, we can imagine group transformations of it, and that these transformations then have consequences for the observable components. In the spin case and in the case with the triangle in the sphere, we can think of rotations. In the treatments of a patient-example it is meaningful to study scale invariance: $(\tau_1, \tau_2) \to (b\tau_1, b\tau_2)$.

So for such a c-variable I cannot ask every question I want, but I ask a question about a maximally observable component, and the answer is what I say defines a state. Even though I start with a mental picture involving unobservables, the state is defined in terms of an observable quantity.

My opinion is that we are close here to the overall goal of finding a link between statistics and quantum mechanics; the technical details concerning quantum mechanics are spelled out in Section 5.4 below. But I also mean that we then must be willing to change focus sometimes in traditional theoretical statistics. This I will come back to in Chapter 7.

In theoretical statistics there seems to be almost complete separation between experimental design theory and inference theory. In practical experiments the two are linked closely together. In fact, a formulation of the experimental question should almost always be an important part of the conclusion, and in any useful investigation in biology and medicine, say, it always is. Thus here also, in my opinion the conclusion should, in all good experiments, be stated as a question plus an answer. Also, all practical experiments are performed in a context. In a quantummechanical setting this means that the choice of question also is done in some context. This aspect will be discussed in the last section of this Chapter.

Statisticians that are close to applications, sometimes think in terms of mental quantities that are close to my c-variables, for instance Searle in his book [183] bases his treatment to a large extent on unestimable linear parameters, motivated by the fact that this gives a nicer mental picture when you want to consider several models at the same time.

Large parts of this chapter are built upon Helland [113], published in a statistical journal and Helland [116], published in a journal for theoretical physics. In a recent series of papers, Wetterich [202–204] has given other arguments for our common thesis that quantum mechanics can be derived from classical statistics.

5.2   The Hilbert Spaces of a Given Experiment

Summary: The definition of a c-system is used to construct a Hilbert space for each single experiment that can be chosen in some given setting. To make the discussion here reasonably self-contained, the mathematical concepts of Hilbert space and L^2-space are also defined.

As I continue this chapter, I will repeat the assumptions of Section 4.6 and Section 4.7 as I need them, and then comment further upon the definition and the axioms and use them in the mathematical development. Since we are approaching the formal apparatus of quantum mechanics, this and the following sections will of necessity be somewhat mathematical. I will try to stress what I feel is the essence of the mathematics as we go along, however. Those who want to, can concentrate on the main results.

I start by repeating for completeness the definition of a c-system, the general framework for a hypothetical set of experiments, either microscopic or macroscopic.

Definition 4.6.1.   *A c-system is a closed system for which there exists a c-variable space Φ whose elements $\phi \in \Phi$ are not estimable relative to any experiment,* *that is, they are inaccessible. There is a transformation group G defined on Φ. The space Φ is locally compact, and the group G is transitive on Φ. The invariant measure on Φ under G is called ν.*

Furthermore, there exists a set $\mathcal{A}$ of potential experiments $\mathbf{E}^a; a \in \mathcal{A}$ on this system. For each $a \in \mathcal{A}$ there is a maximal estimable parameter λ^a.

When λ^a is discrete, there exists a perfect experiment for λ^a where measurement noise can be disregarded and where the resulting (pure) states are given by statements of the form $\lambda^a = \lambda_k^a$.

The concept of an inaccessible c-variable should now be well known. We must imagine that the space Φ is so large that every conceivable estimable parameter is a function of ϕ. That the c-variable space is locally compact, is a weak technical assumption. It is also a weak assumption that there exists a transformation group G on this space. A simple example where there is a natural such group is Example 1.3.4 with the expected survival time for two hypothetical treatments of the same patient. In this example G can be the time scale group.

The existence of an invariant measure ν on Φ is proved in Theorem A.1 in Appendix A.2.2 under weak assumptions.

A main assumption is that G is transitive on Φ. This means that the transformation group has only one orbit on the space Φ: Starting at any point $\phi_0 \in \Phi$ we can reach every other point in the space by an appropriate group transformation. This property can be ensured by taking G large enough. It may not hold if our system consists of completely separated subsystems, but it does as a rule hold for particle systems under permutational symmetry, and for entangled systems, which we will discuss later.

The assumption that there exist experiments $\mathbf{E}^a$ as stated is a relatively weak one. It only says that the c-variable ϕ can be divided into parameters $\lambda^a = \lambda^a(\phi)$ connected to experiments. Our choice of experiment a is essential; it corresponds to our choice of focus in our investigation of the system in question.

The parameter λ^a is the important one, since it is estimable in the selected experiment $\mathbf{E}^a$. Later, we will look upon the λ^a's as state variables, and make assumptions to that end. The requirement that λ^a is maximal is crucial. Mathematically it means that there exists no other estimable paramater μ^a such that λ^a is a proper function of it.

One can imagine that λ^a consists of several components, and the question: What is the value of λ^a? will then consist of several partial questions. The requirement of maximality then means that this list of partial questions is as long as it can possibly be, given the physical system in question, and given that the questions are independent.

Recall from Definition 4.7.1 that the subgroup G^a of G was defined as the maximal subgroup with respect to which $\lambda^a(\cdot)$ is a permissible function. From Lemma 3.4.1 this subgroup always exists.

Example 5.2.1. In the spin example in Section 4.8 we could also have used an equivalent alternative definition: Take G^a as the maximal subgroup with respect to which $\theta^a(\cdot)$ is a permissible function, where θ is the intermediate parameter defined there. Again this would have led to the following subgroup of the rotation group: G^a is the group of rotations around a together with a single rotation around an axis perpendicular to a. Then the reduction $\theta^a \mapsto \lambda^a$ is a reduction to an orbit of the group G^a. These orbits are all of the form $\lambda^a = \pm k$ for some $k \geq 0$; in particular the set $\lambda^a = \pm 1$ forms an orbit for the group G^a as acting upon θ^a.

This example is related to the kind of model reduction that we argued for in Chapter 3 and will give further examples in the Chapters 7 and 8: Every sensible model reduction should be to an orbit or to a set of orbits of the relevant group.

The definitions above are what we need to construct the Hilbert space for a single experiment. The concept of a Hilbert space has already been mentioned several times, and it was used in the three quantum rules in Chapter 4. Now it is time to give the formal definitions.

Definition 5.2.1. *A Hilbert space is a vector space which is closed under the norm $\| v \| = \sqrt{(v,v)}$ formed from a scalar product (v_1, v_2).*

This is a standard mathematical definition. A vector space is a space which for every element v_1 and v_2 contains all linear combinations $\alpha_1 v_1 + \alpha_2 v_2$. In this book, as is common in quantum mechanics, we will consider complex vector spaces, i.e., the scalars α_1 and α_2 above may be complex numbers. The Hilbert space may have a finite or an infinite dimension. To appreciate the concepts involved in Definition 5.2.1, we need some further definitions:

Definition 5.2.2. *A scalar product (v_1, v_2) in a vector space is a complex function of the two vectors which satisfies*

$$(v_1, \alpha_2 v_2 + \alpha_4 v_4) = \alpha_2(v_1, v_2) + \alpha_4(v_1, v_4).$$

$$(v_1, v_2)^* = (v_2, v_1),$$

$$(v, v) \geq 0,$$

$$(u, v) = 0 \text{ for all } v \text{ if and only if } u = 0.$$

Here * denotes complex conjugation. The simplest case of a Hilbert space is an n-dimensional complex vector space, that is, a space which has a basis $\{e_i; i = 1, ..., n\}$

such that every vector is a linear combination of these basis vectors. Then, if $v_1 = \sum_{i=1}^{n} \beta_i e_i$ and $v_2 = \sum_{i=1}^{n} \gamma_i e_i$, we can define the scalar product

$$(v_1, v_2) = \sum_{i=1}^{n} \beta_i^* \gamma_i.$$

Definition 5.2.3. *The vector space is closed in the norm $\| v \|$ if $\{v_n\}$ converges to some vector v in the vector space whenever $\| v_m - v_n \|$ tends to 0 as m and n tend to infinity. This implies that $\{v_n\}$ converges to v whenever $\| v_n - v \|$ tends to 0.*

For our purpose in the sequence it is important to note that a space of functions can easily be a vector space. We first need to define another standard mathematical concept:

Definition 5.2.4. $L^2(\Phi, \nu)$ *is the Hilbert space of complex functions f on Φ with scalar product*

$$(f_1, f_2) = \int f_1^*(\phi) f_2(\phi) \nu(d\phi). \tag{5.2}$$

It is a straightforward exercise to show that $L^2(\Phi, \nu)$ satisfies the definition of a Hilbert space. Our construction of a Hilbert space for quantum mechanics will take this space as a point of departure. We first introduce the Hilbert space corresponding to a particular experiment.

Definition 5.2.5. *Let $\mathbf{L} = L^2(\Phi, \nu)$, and let the Hilbert space $\mathbf{L}^a$ for experiment a be the set of functions f in $\mathbf{L}$ which can be written in the form $f(\phi) = \tilde{f}(\lambda^a(\phi))$.*

Of all the definitions of this section, this may be the most important to remember in the following: For each a, the space $\mathbf{L}^a$ is just the space of functions of the simple form

$$f(\phi) = \tilde{f}(\lambda^a(\phi)).$$

It is clear that $\mathbf{L}^a$ is a linear space: The linear combination of two functions of λ^a will be a function of λ^a. It is also a closed subspace of $\mathbf{L}$: This means that any converging sequence of functions in $\mathbf{L}^a$ will converge towards a function in $\mathbf{L}^a$. It is a standard result that a closed subspace of a Hilbert space is a Hilbert space. Hence $\mathbf{L}^a$ is a Hilbert space.

In the next section we will make a simple transformation of $\mathbf{L}^a$ to make an equivalent Hilbert space $\mathbf{H}^a$.

In Helland [113] a more cumbersome definition of $\mathbf{H}^a$ was made, where the sample space was taken into account. It was proved there ([113], Proposition 3) that for the case of a perfect experiment this definition coincides with our Definition 5.2.5. In our further development of the Hilbert space here, we will limit ourself to perfect experiments, and come back to real experiments later.

Quantum variables like position and momentum, variables which are not discrete, require a more elaborate formalism than the ordinary quantum Hilbert space formalism if we want to be precise. To start with we will therefore limit ourselves to discrete parameters λ^a, and then the existence of a perfect experiment is clear by the last part of Definition 4.6.1.

The formulation of a state in this part of Definition 4.6.1 corresponds to our earlier definition of a state as a question plus an answer. One of our immediate goals is to to show that this can be made equivalent to the quantum-mechanical definition of a state as given by a unit vector in the common Hilbert space.

5.3   The Common Hilbert Space

Summary: The concept of unitary operator is defined, and some basic aspects of group representation theory are discussed. Under Axiom 1 it is shown that the Hilbert spaces of the single experiments can be unitarily connected, and that this can be used to construct a common Hilbert space, a basic postulate of conventional quantum mechanics. Under Axiom 2 it is shown that this common Hilbert space is an invariant space for a representation of the basic group.

Our task in this section is to tie the spaces $\mathbf{L}^a$ together. The situation we have in mind is one where the parameter spaces of the different experiments have a similar structure. Then it is not unreasonable to assume that they can be transformed over to each other by some element of the basic group G. This will not give the most general case of the quantum-mechanical formalism, but gives a treatment which includes qubits, higher spins, several particles with spin and the most important cases of entanglement, a phenomenon which is much discussed in the quantum-mechanical literature.

What we will do in this section is to build one common Hilbert space from the separate Hilbert spaces of the previous section. This will then be our starting point towards the formalism of quantum theory, which is built upon one single abstract Hilbert space. Note that in order to find a valid starting point for this task, it is enough, as we do, to construct one concrete Hilbert space. This is because the abstract Hilbert space of quantum mechanics is separable, a property shared by our Hilbert space constructed below, and there is a well-known mathematical proposition saying that all separable Hilbert spaces of the same dimension are unitarily equivalent, that is, there is a unitary invertible linear transformation between each pair of them. This can be seen by considering the bases of the different Hilbert spaces, and let the unitary transformation between two spaces be defined through a transformation between the bases. The definitions follow.

Definition 5.3.1. *a) A Hilbert space* **H** *is called separable if there is a countable basis* $\{e_i\}$ *in* **H** *such that each vector* $v \in$ **H** *can be written in the form of a convergent sum*

$$v = \sum_i \beta_i e_i.$$

b) A linear operator U *in the Hilbert space* **H** *is called unitary if*

$$(Uv_1, v_2) = (v_1, U^{-1}v_2)$$

for all $v_1, v_2 \in$ **H**.

In the finite-dimensional case a unitary operator will be a unitary complex-valued matrix. A square matrix U is unitary if we get its inverse by transposing and then taking its complex conjugate.

A direct verification gives: The product of two unitary operators is unitary. The inverse of a unitary operator is unitary.

Let W be a fixed unitary matrix. Then we define $\mathbf{H}^a = W\mathbf{L}^a$ as an alternative space characterizing the experiment with parameter λ^a. It is easy to see that $\mathbf{H}^a$ also is a Hilbert space.

Since we now will go from a set of axioms involving symmetry and groups to a vector space formulation, it turns out to be useful to represent group elements as matrices, or more generally, as operators on some vector space. This leads to the well known mathematical area called *group representation theory*. To give an analogy which should not be taken too literally: In a similar way as characteristic functions are used in probability theory to characterise probability distributions, representations of group elements as operators on some vector space may be used to characterise the relevant groups.

Definition 5.3.2. *A representation* V *of a group* G *on a vector space* **L** *is defined as a homomorphism from* G *to the space of operators on* **L**, *i.e.. a group of operators which satisfy*

$$V(gh) = V(g)V(h) \text{ whenever } g, h \in G \tag{5.3}$$

The representation is called unitary if all the operators $V(g)$ *are unitary.*

With the basic group G defined as in Definition 4.6.1 repeated in Section 5.1, the simplest case is the following.

Definition 5.3.3. *The right regular representation* U *of* G *on* $\mathbf{L} = L^2(\Phi, \nu)$ *is defined by*

$$U(g)f(\phi) = f(\phi g). \tag{5.4}$$

for all $f, g.\phi$.

It is straightforward to verify that each $U(g)$ as defined in 5.3.3 is a linear operator, and that the homomorphism property (5.3) holds:

$$U(gh)f(\phi) = f(\phi gh) = U(g)f(\phi h) = U(g)U(h)f(\phi).$$

The second equality here is justified by taking $r(\phi) = f(\phi h)$ and then $U(g)r(\phi) = r(\phi g)$.

Lemma 5.3.1. *The right regular representation is a unitary representation with respect to the scalar product defined in (5.2).*

Proof.

$$(U(g)f_1, f_2) = \int f_1^*(\phi g)f_2(\phi)\nu(d\phi) = \int f_1^*(\phi)f_2(\phi g^{-1})\nu(d\phi) = (f_1, U(g)^{-1}f_2).$$

Here the invariance property of the measure ν was used, taking Definition 4.6.1 as the point of departure. Note that in all the proofs here, as in the introduction of the right-invariant measure in Chapter 3, our convention of placing the group element to the right of the element to be transformed, leads to the simplest formulation.

Going from G to the subgroup G^a, we get a subgroup U^a of U, again a group of unitary operators, the regular representation of the group G^a.

Proposition 5.3.1. *a) The Hilbert space $\mathbf{L}^a$ is an invariant space for the representation U^a:*

$$U^a(g)f \in \mathbf{L}^a \text{ whenever } g \in G^a, \ f \in \mathbf{L}^a. \tag{5.5}$$

b) The Hilbert space $\mathbf{H}^a = W\mathbf{L}^a$ is an invariant space for the representation $V^a = WU^aW^\dagger$.

Again the verification is straightforward: If $f \in \mathbf{L}^a$, then $f(\phi) = \tilde{f}(\lambda^a(\phi))$. Therefore, for $g \in G^a$:

$$U^a(g)f(\phi) = f(\phi g) = \tilde{f}(\lambda^a(\phi g)),$$

so $U^a(g)f \in \mathbf{L}^a$.

Invariant spaces turn out to be important in group representation theory. A summary of this theory is given in Appendix A.2.4.

After this introduction, it is time to repeat the axioms from Section 4.7. The following one is crucial:

Axiom 1. *For each pair of experiments $a, b \in \mathcal{A}$ there is an element g_{ab} of the basic group G which induces a correspondence between the respective parameters:*

$$\lambda^b = \lambda^a g_{ab} \text{ or } \lambda^b(\phi) = \lambda^a(\phi g_{ab}). \tag{5.6}$$

This assumption is fairly strong, and it makes the task of connecting the spaces really simple. On the other hand, it seems to be satisfied in concrete cases. The same assumption will be needed in Section 5.5.

A simple example is the triangle in a sphere case discussed in Section 4.5, where the group G is either the rotation group or the permutation group. Any rotation/permutation which takes window a to window b can be used as g_{ab}.

In the electron spin case Φ was a space of vectors, G was the rotation group together with changes of scale. Then (5.6) holds if g_{ab} is any rotation transforming a to b.

If (5.6) holds for transformations on some component spaces, it also holds for the cartesian product of these spaces when the relevant cartesian product of groups are used.

Another interesting relation is connected to Axiom 1 in the following way: The relationship (5.6) implies that one must have $\lambda^b g^b = \lambda^a g^a g_{ab}$ for some $g^b \in G^b$. Hence it follows that $\lambda^a g_{ab} g^b = \lambda^a g^a g_{ab}$, so, since the two group elements below act upon λ^a in the same way:

Lemma 5.3.2. *It follows from Axiom 1 that* $g^a = g_{ab} g^b g_{ab}^{-1}$.

One can give many examples of group transformations where $g^a = g_{ab} g^b g_{ab}^{-1}$ holds in general, giving a homomorphism between the groups G^a and G^b. This relation

$$g^a = g_{ab} g^b g_{ba}$$

is what mathematicians call an inner homomorphism between group elements, or really an isomorphism. An isomorphism means that essentially the same group is acting upon both spaces $\Lambda^a = \{\lambda^a\}$ and $\Lambda^b = \{\lambda^b\}$, and often in such cases the same group element symbol is used. We will use different symbols, however, because the actions are related to different experiments.

Axiom 1 will be crucial in connecting the Hilbert spaces $\mathbf{H}^a$ for the different experiments. First, from the construction of the Hilbert spaces, $\mathbf{L}^a$ is a space of functions of $\lambda^a(\phi)$, and $\mathbf{L}^b$ is a space of functions of $\lambda^b(\phi)$. Furthermore, the spaces are constructed in the same way. Specifically, if $f^a(\phi) = \tilde{f}(\lambda^a(\phi))$ and $f^b(\phi) = \tilde{f}(\lambda^b(\phi))$, then by (5.6) we have

$$f^b(\phi) = f^a(\phi g_{ab}) = U(g_{ab}) f^a(\phi). \tag{5.7}$$

Here U is the right regular representation of the group G.

Now introduce an assumption which ensures that the set of transformations on the parameters λ^a of the perfect experiments $\mathbf{E}^a$ generate all the transformations in the transitive group G on the c-variable space Φ.

Axiom 2. *The reduced groups $G^a, G^b, \ldots$ generate the whole group G.*

From this we get:

Theorem 5.3.1. *For any choice of W, the fixed space $\mathbf{H}^a = W\mathbf{L}^a$ is an invariant space for some abstract representation (possibly multivalued) $V(g)$ of the whole group G.*

Proof.
It follows from Proposition 5.3.1 that $\mathbf{L}^a$ is an invariant space for the group G^a under the right regular representation U. This follows directly from the definitions. It follows that $\mathbf{H}^a = W_0\mathbf{L}^a$ is an invariant space for G^a under the representation $V_0(g) = W_0 U(g) W_0^\dagger$.

This can now be extended. Look at the product $g_1 g_2 g_3$, where $g_1 \in G^a$, $g_2 \in G^b$ and $g_3 \in G^c$. We can define a map from such elements to operators on $\mathbf{H}^a$ by

$$V(g_1 g_2 g_3) = V_0(g_1)(V_0(g_{ab})W_0 U(g_2)W_0^\dagger V_0(g_{ba}))(V_0(g_{ac})W_0 U(g_3)W_0^\dagger V_0(g_{ca})).$$

$$(5.8)$$

With a similar definition for all elements from the subgroups, we verify for instance that $V(g_1 g_2')V(g_2'' g_3) = V(g_1 g_2 g_3)$ when $g_2' \in G^b$, $g_2'' \in G^b$ and thus $g_2 = g_2' g_2'' \in G^b$. Also, $V(g_1 g_2 g_3) = V(g_1)V(g_2)V(g_3)$ when $g_1 \in G^a$, $g_2 \in G^b$ and $g_3 \in G^c$.

Let now g and h be any two elements in G such that g can be written as a product of elements from the subgroups G^a, G^b and G^c, and similarly for h. (The proof is similar in other cases.) It follows that $V(gh) = V(g)V(h)$ on these elements, since the last factor of g and the first factor of h either must belong to the same subgroup or to different subgroups, so the statement follows from the previous 2 statements.

In this way we see that V is a representation on the set of finite products of elements from the subgroups, and since by Axiom 2 these products generate G, it is a representation of G.

This is true for all choices of W_0. In particular, also, $\mathbf{H}^a$ is an invariant space of the representation $V(g)$ of the subgroup G^a. The aim of the rest of this proof is to show that this can be extended to all of G by a particular choice of $V(g)$.

First, let W_b be the operator which diagonalizes $V_0(g_{ab}) = W_0 U(g_{ab})W_0^\dagger$ on the space $\mathbf{H}^a$. Then $W_b V_0(g_{ab})W_b^\dagger \mathbf{H}^a = \mathbf{H}^a$. Let $V_1(g) = W_b V_0(g)W_b^\dagger$. Then $V_1(g_{ab})\mathbf{H}^a = \mathbf{H}^a$, and $V_1(g^a)\mathbf{H}^a = \mathbf{H}^a$ for $g^a \in G^a$. For $g_1 \in G^a$ and $g_2 \in G^b$ we have

$$V_1(g_1 g_2)\mathbf{H}^a = V_1(g_1 g_{ab}^{-1} g^a g_{ab})\mathbf{H}^a = \mathbf{H}^a.$$

Proceed successively with similar constructions for $G^c, G^d, \ldots$, and find that successive operators leave $\mathbf{H}^a$ invariant under all products of group elements from the subgroups. Use again that these products generate G, and finally perform a limiting operation, noting that the right regular representation $U(g)$ is a continuous function of g for a suitable topology on G.

Since different representations of g as a product may give different solutions, we have to include the possibility that V may be multivalued. What we need, however, is the existence of one representation V.

Choice of W and hence of the common Hilbert space: Now keep $\mathbf{H}^a$ fixed, but be free to change the operator W_0 which fixes the relation to the other Hilbert spaces. Concretely, there is a group representation having $\mathbf{H}^a$ as an invariant space. By Theorem 5.3.1, we can choose W such that $V(g) = WU(g)W^\dagger$ is this representation.

Theorem 5.3.2. *Then from* $\mathbf{H}^b = V(g_{ab})\mathbf{H}^a$ *we have* $\mathbf{H}^a = \mathbf{H}^b = \mathbf{H}^c = ...$, *and this can be taken as the quantum-mechanical space* $\mathbf{H}$.

As an example, the two-dimensional Hilbert space of a particle with spin is an irreducible invariant space for the rotation group with a two-valued representation $SU(2)$ - upon the set of unitary 2-dimensional matrices with determinant $+1$. Such a representation determines to a large extent $\mathbf{H}$. In general, the requirement that $\mathbf{H}$ should be a representation space for G may put a constraint on the dimension of $\mathbf{H}$.

The construction above gives a concrete representation of the quantum-mechanical Hilbert space. Since all separable Hilbert spaces of the same dimension are unitarily equivalent, other representations - or just an abstract representation - may be used in practice. This is sufficient to give the Born formula as proved below, and through this the ordinary quantum formalism. But the concrete representation facilitates interpretation.

5.4 States and State Variables

Summary: The basic elements of conventional quantum theory are constructed from our setting: First the state vectors of the Hilbert space are constructed and shown each to be in one-to-one correspondence with a focused question concerning the value of a maximal parameter together with a crisp answer of this question. Next we construct an operator corresponding to the parameter which has the state vector as an eigenvector and the crisp parameter value as an eigenvalue. It is shown that in general it is not true that all unit vectors in the Hilbert space can be given such an interpretation: Among other things superselection rules may exclude certain vectors. Finally I look briefly at the case of a compound parameter and correspondingly a Hilbert space which is a direct product of single spaces.

By what has just now been proved, for each a there is a Hilbert space $\mathbf{H}^a$ of functions of λ^a, and these can be put in unitary correspondence with a common Hilbert space $\mathbf{H}$. In this section and the next few we shall make an assumption which is common in elementary quantum mechanics, but which is very restrictive from a statistical point of view, namely the assumption of discreteness:

Axiom 3. *Each parameter λ^a assumes a finite or denumerable set of values.*

This is consistent with Definition 4.6.1 and the existence of a perfect experiment for λ^a. In the finite case Axiom 3 implies that the group G^a acts upon λ^a as a group of permutations.

The following result may be of some interest:

Lemma 5.4.1. *These discrete values can be arranged such that each $\lambda_k^a = \lambda_k$ is the same for all a ($k=1,2,...$).*

Proof.

By Axiom 1

$$\{\phi : \lambda^b(\phi) = \lambda_k^b\} = \{\phi : \lambda^a(\phi g_{ab}) = \lambda_k^b\} = \{\phi : \lambda^a(\phi) = \lambda_k^b\} g_{ba}.$$

The sets in brackets on the left-hand side here are disjoint with union Φ. But then the sets on the right-hand side are disjoint with union $\Phi g_{ab} = \Phi$, and this implies that $\{\lambda_k^b\}$ gives all possible values of λ^a.

In the finite case we get that G^a, as acting upon λ^a, is a group of permutations, and that the corresponding invariant measure is the counting measure.

In spite of Lemma 5.4.1 - since in a statistical model a parameter always can be changed to any one-to-one function of it - I will keep the notation λ_k^a in order to have the most general treatment.

I will now consider my claim that a quantum-mechanical state can be associated in a unique way by a focused question together with a crisp answer. This will be taken as a definition, and then I will show that this definition makes sense in a quantum-mechanical setting.

Look first at the electron spin case; there one can get a direct characterisation: Let the question be about the spin in direction a, and let the answer be +1, then define as an abbreviation for this state, the 3-vector $u = a$. If the answer is -1, let the state be characterised by the 3-vector $u = -a$. This is consistent, since the latter state also can be result +1 from a chosen measurement direction $-a$. Thus from my definition, the state can be characterised by a 3-vector u. For further development, see Proposition 5.6.3 below.

In the general Hilbert space every normalized vector is the eigenvector of some operator T, and if this operator corresponds to a meaningful physical variable, then one has a question and answer situation. There turns out to be some uniqueness here under the stated assumptions. I start with the component Hilbert space.

Definition 5.4.1. *In the space $\mathbf{L}^a$ let $f_k^a(\phi)$ be defined as the trivial function*

$$f_k^a(\phi) = 1 \text{ when } \lambda^a(\phi) = \lambda_k^a, \tag{5.9}$$

otherwise $f_k^a(\phi) = 0$. These functions are orthogonal basis functions for $\mathbf{L}^a$. The functions f_k^a are also eigenfunctions of the operator S^a defined by

$$S^a f(\phi) = \lambda^a(\phi) f(\phi). \tag{5.10}$$

Since λ^a is a maximal parameter by Definition 4.6.1, the eigenfunctions above are non-degenerate.

Now we turn to the common Hilbert space **H**. Recall from Theorem 5.3.2 that **H** is an invariant space under the representation $V(g)$ of the group G. It may perhaps be illuminating to mention that in the case of a compact group G, it is a standard result from group representation theory [19] that every representation is a subrepresentation of the right regular representation. Hence in this case, each $V(g)$ acts as the component of $U(g)$ along the space **H**.

Also recall that $U(g_{ac})f_k^a$ belongs to $\mathbf{H} = \mathbf{H}^c$; in fact it is one of the functions f_j^c since $f_k^a(\phi g_{ac}) = I(\lambda^a(\phi g_{ac}) = \lambda_k^a) = f_j^c(\phi)$ for some j by Lemma 5.4.1.

Definition 5.4.2. *In the space* $\mathbf{H} = \mathbf{H}^a = W\mathbf{L}^a$ *the functions* f_k^a *correspond to the vectors* v_k^a *defined by*

$$v_k^a = W f_k^a . \tag{5.11}$$

By definition v_k^a is the state vector corresponding to the question: *What is the value of* λ^a? *together with the answer:* $\lambda^a = \lambda_k^a$.

Remark 1. It is convenient here, as in ordinary quantum mechanics, to give the same interpretation to unit vectors v_k^a which are the same except for a constant phase factor. This can be done by redefining f_k^a as any function of the form $f(\phi) = \exp(i\gamma)I(\lambda^a(\phi) = \lambda_k^a)$. The interpretation of any such f_k^a is the same.

Remark 2. For each a the vectors $\{v_k^a\}$ form an orthogonal basis for the space **H**. In the p-dimensional case, if the measure ν is normalised such that $\nu(\{\phi : \lambda^a(\phi)\}) = 1/p$, the basis is orthonormal.

Remark 3. In general, the vectors v_k^a are eigenvectors of the selfadjoint operator

$$T^a = W S^a W^\dagger \tag{5.12}$$

with eigenvalues λ_k^a. Hence we have the result:

Theorem 5.4.1. *For each a there is an operator T^a on the Hilbert space **H** which corresponds to the perfect experiment $\mathbf{E}^a$ with the parameter λ^a. These operators have eigenvectors v_k^a with eigenvales λ_k^a.*

Note that by the unitary transformation, the vectors v_k^a are unit vectors when the f_k^a are. An eigenvector v_k^a by definition represents the statement that the parameter λ^a has been measured with a perfect measurement that has given the value λ_k^a. In fact the unit vectors in **H** correspond in a unique way to such a question-and-answer pair if there is a correspondence at all:

Theorem 5.4.2. *a) Assume that two vectors in* $\mathbf{H}$ *satisfy* $v_i^a = v_j^b$, *where* v_i^a *corresponds to* $\lambda^a = \lambda_i^a$ *for a perfect experiment* $\mathbf{E}^a$ *and* v_j^b *corresponds to* $\lambda^b = \lambda_j^b$ *for a perfect experiment* $\mathbf{E}^b$. *Then there is a one-to-one function* F *such that* $\lambda^b = F(\lambda^a)$ *and* $\lambda_j^b = F(\lambda_i^a)$.

b) Each v_k^a *corresponds to only one* (λ^a, λ_k^a) *pair except possibly for a simultaneous one-to-one transformation of the latter.*

The proof of Theorem 5.4.2 is given in Appendix A.3.2.

Above we started with indicator functions f_k^a in $\mathbf{H}^a$ as eigenfunctions of S^a with eigenvalues λ_k^a. These indicator functions can be transformed in the natural way by the group elements $g^a \in G^a$, the subgroup of transformations of Λ^a, and by the corresponding regular unitary representation operators $U(g^a)$.

Proposition 5.4.1. *If* f_k^a *is an eigenfunction for* S^a *with eigenvalue* λ_k^a, *then* $U(g^a)f_k^a$ *is an eigenfunction for* S^a *with eigenvalue* $\lambda_k^a g^a$.

The proof uses the fact that $U(g^a)f_k^a(\phi) = f_k^a(\phi g^a)$, so $S^a f_k^a = \lambda_k^a f_k^a$ implies $S^a f_k^a(\phi g^a) = \lambda_k^a f_k^a(\phi g^a)$, which implies $U^{-1}(g^a)S^a U(g^a)f_k^a(\phi) = \lambda_k^a g^a f_k^a(\phi)$.

For some cases, like with the electron spin, it can be demonstrated directly that that the set of unit vectors v_k^a as k and a vary constitute all the unit vectors in $\mathbf{H}$, if we include a trivial phase factor. In general it is not true that all unit vectors of $\mathbf{H}$ can be given such an interpretation. Among other things, one has to take into account what is called superselection rules: For an absolutely conserved quantity μ like charge or mass, the linear combinations of eigenvectors corresponding to different eigenvalues of the operator associated to μ are not possible state vectors. Superselection rules are well known among physicists, but they are not always stressed in textbooks in quantum mechanics.

It is a general theorem from representation theory [19] that every finitedimensional invariant space can be decomposed into a direct sum of irreducible invariant spaces. Recall that an invariant space is irreducible if it contains no further invariant space. Thus we have

$$\mathbf{H}^a = \mathbf{H}_1^a \oplus \mathbf{H}_2^a \oplus \ldots,$$

where the $\mathbf{H}_i^a$ are irreducible representation spaces under $U^a(g)$, and where the symbol $\oplus$ denotes direct sum: Here the superpositon of orthogonal spaces.

By doing this in both the experiments $\mathbf{E}^a$ and $\mathbf{E}^b$ we see that we can assume that the relation

$$U^a(g^a)U(g_{ab}) = U(g_{ab})U^b(g^b)$$

holds separately between irreducible spaces on each side. If $\mathbf{H}_1^a$ is a space of functions of $\lambda^a(\phi)$, irreducible under U^a, then

$$U(g_{ab})\mathbf{H}_1^a$$

must be a space of functions of $\lambda^b(\phi)$, irreducible under U^b. This gives unitary relations between pairs of irreducible spaces, one for each experiment, and the space $\mathbf{H}^b$ has a corresponding decomposition $\mathbf{H}_1^b \oplus \mathbf{H}_2^b \oplus \dots$.

It follows also that $\mathbf{H} = \mathbf{H}_1 \oplus \mathbf{H}_2 \oplus \dots$ conformably, and that there is a unitary connection $\mathbf{H}_i^a = R_i^a \mathbf{H}_i$.

Summarizing the above discussion we have:

Theorem 5.4.3. *The basic Hilbert space* $\mathbf{H}$ *can be decomposed as* $\mathbf{H} = \mathbf{H}_1 \oplus \mathbf{H}_2 \oplus \dots$, *where each* $\mathbf{H}_i$ *is an irreducible invariant space under the representation* $\{V(g)\}$ *of the group* G. *Each part corresponds to a fixed value of one or several quantities that are conserved under all experiments.*

Again the question is if all unit vectors of the spaces $\mathbf{H}_i$ are possible state vectors. This will not be true in general, but will be proved in Section 5.6 to hold in the electron spin case. In the general case it might be that the Hilbert spaces $\mathbf{H}_i$ contain other orthonormal sets of basis functions than the v_k^a constructed above. Each such set of basis functions together with a corresponding set of eigenvalues lead to an operator T. Thus, if this operator has some physical meaning, then in the same way as above, we can say that each eigenvector corresponds to some question plus a crisp answer. But in this case, the question is not of the simple form: What is the value of λ^a? for any of the parameters λ^a considered earlier.

To conclude: In simple terms a state is characterised by the fact that a (maximal) perfect measurement is performed, and this has lead to some value of the corresponding maximal parameter. Concretely: A perfect experiment $a \in \mathcal{A}$ has led us to consider the Hilbert space $\mathbf{H}^a$, and the result $\lambda^a = \lambda_k^a$ is exactly characterised by the indicator function f_k^a. Translated to the $\mathbf{H}$-space, the state given by the information $\lambda^a = \lambda_k^a$ is then characterised by the vector v_k^a.

A word should be inserted on my use of parameters here. Usually in quantum mechanics a state of the spin of an electron is specified by saying that, say, the spin component in direction a *is equal to* $+1$. In my formulation I say that the parameter λ^a has been measured to be equal to $+1$. But I stress then that the measurement then should be in terms of a perfect experiment. And saying that a component is equal to $+1$ and saying that the component has been measured by a perfect experiment to be equal to $+1$, amounts to nearly the same thing. And it amounts to exactly the same thing if we adhere to an epistemological interpretation of quantum mechanics, which I do throughout this book.

This is consistent with the well known quantum-mechanical interpretation of a state vector. In my treatment, this interpretation of a state as a question-answer pair is crucial.

The operator T^a may be written

$$T^a = \sum_k \lambda_k v_k^a v_k^{a\dagger}. \tag{5.13}$$

Roughly again, $v_k^{a\dagger}$ is a row vector, the transpose and complex conjugate of the coloumn vector v_k^a. This interpretation is precise in the finite-dimensional case. In general the definition is given by $v_k^{a\dagger}u = (v_k^a, u)$ for all u.

The operators T^a are selfadjoint, meaning that $T^{a\dagger} = T^a$ in a similar interpretation. This implies that the operators have real eigenvalues λ_k^a. They also satisfy the trivial relation $v_k^{a\dagger}T^a v_k^a = \lambda_k^a$.

Using the results of this section to construct the joint state vector for a system consisting of several partial systems, with symmetries only within the partial systems, one follows the recipe $v_{k_1 k_2 k_3}^{a_1 a_2 a_3} = v_{k_1}^{a_1} \otimes v_{k_2}^{a_2} \otimes v_{k_3}^{a_3}$, where it is assumed that system k is in state $\lambda^{a_r} = \lambda_{k_r}$ for $r = 1, 2, 3$, and where $\otimes$ denotes the direct product of vectors. By time development under interaction, as described by the Schrödinger equation, or by other means, other, entangled, multicomponent states will occur. This will be further discussed in Chapter 6.

5.5   The Born Formula

Summary: *The purpose of this section is to prove Born's formula from my assumptions. Born's formula is the celebrated quantummechanical result saying how, knowing the state vectors, we can compute the transition probability from one state to another.*

We have now obtained a statistical interpretation of the quantum-mechanical Hilbert space: Unit vectors in that space can be equivalently characterised as question-answer pairs, and, furthermore, the Hilbert space is invariant under a suitable representation of the basic group G.

To complete deriving the formalism of quantum mechanics from the statistical parameter approach the most important task left is to arrive at the Born formula, which gives the probability of transition from one state to another. The fact that such a formula exists is amazing, and must be seen as a result of the symmetry of the situation together with the fact that each parameter λ^a is maximal and together with the limitation imposed by the Hilbert space. Even though I use a different approach, my own result is related to recent attempts to link the formula to general decision theory: An interesting development which goes in this direction was recently initiated by Deutsch [60]. The approach of Deutsch has been criticised by Finkelstein [81], by Barnum et al. [16] and by Gill [95], who gave a constructive set of arguments using three reasonable assumptions.

In this section I will concentrate on the case with one irreducible component in the Hilbert space, i.e., I will neglect superselection rules. This is really no limitation, since transitions between different components are impossible.

What I am going to prove is a result connecting two different perfect experiments in the same system. Assume that we know from the first perfect experiment that

$\lambda^a = \lambda^a_k$. Next assume that we perform another perfect experiment $b \in \mathcal{A}$. In both cases, the notion of perfect measurement means that measurement error can be neglected. More realistic experiments are treated in Section 5.6 below. In the perfect case it turns out that we can find a formula for

$$P(\lambda^b = \lambda^b_i | \lambda^a = \lambda^a_k)$$

which only depends upon the state vectors v^a_k and v^b_i.

This formula has a large number of important consequences in quantum mechanics, and, as already said, it can be argued for in different ways. I will prove it from the following

Axiom 4. *(i) The transition probabilities exist in the sense that the probabilities above do not depend upon anything else.*

(ii) The transition probability from $\lambda^a = \lambda^a_k$ in the first perfect experiment to $\lambda^a = \lambda^a_k$ in the second perfect experiment is 1.

(iii) For all a, b, c we have that $\mu(\phi) = \lambda^a(\phi g_{bc})$ is a valid experimental parameter.

(iv) For all a, b, c, i, k we have

$$P(\lambda^b(\phi) = \lambda^b_i | \lambda^a(\phi) = \lambda^a_k) = P(\lambda^b(\phi g_{bc}) = \lambda^b_i | \lambda^a(\phi g_{bc}) = \lambda^a_k).$$

Remarks.

1) Crucial assumptions will also be those implicit in Definition 4.6.1, that a common sample space can be used in all experiments, and Axiom 1. The assumption (i) that the transition probability exists may be related to the assumption that each parameter λ^a is maximal.

2) We have $\lambda^b(\phi g_{bc}) = \lambda^c(\phi)$, so three experimental parameters are included in Axiom 4.

3) In the proof below I need to transform a single experiment by some element of G. The use of the transformation g on the complete sufficient statistics t is then justified by Proposition 3.11.2.

Born's formula is given by:

Theorem 5.5.1. *Under the assumptions above the transition formula is as follows:*

$$P(\lambda^b = \lambda^b_i | \lambda^a = \lambda^a_k) = |v^{a\dagger}_k v^b_i|^2. \tag{5.14}$$

The proof will depend upon a recent variant (Busch [38]; Caves et al. [41]) of a well known mathematical result by Gleason [97]. One advantage of this recent variant is that it also is valid for dimension 2, when the ordinary Gleason Theorem fails.

The Busch-Gleason's theorem. *Consider any Hilbert space* **H**. *Define the set of effects as the set of operators on this Hilbert space with eigenvalues in the interval* $[0, 1]$. *Assume that there is a generalized probability measure* π *on these effects, i.e., a set function satisfying*

$\pi(\mathcal{E}) \geq 0$ *for all* $\mathcal{E}$,

$\pi(\mathcal{I}) = 1$,

$\sum_i \pi(\mathcal{E}_i) = \pi(\mathcal{E})$ *for effects* $\mathcal{E}_i$ *whose sum is an effect* $\mathcal{E}$.

Then π *is necessarily of form* $\pi(\mathcal{E}) = \text{tr}(\rho\mathcal{E})$ *for some positive, selfadjoint, trace 1 operator* ρ.

Remark.

As will be further discussed in the next section, a positive, selfadjoint operator on **H** is an operator of the form

$$\rho = \sum_k \pi_k v_k v_k^\dagger,$$

where $\{v_k\}$ is an orthonormal basis for **H**, and all $\pi_k \geq 0$. This operator has trace 1 if $\sum_k \pi_k = 1$.

The effects involved in the Busch-Gleason theorem turn out to have a rather straightforward statistical interpretation. Look at an experiment b, corresponding to a parameter λ^b which can take the values λ_i^b. Let the result of this experiment be given by a discrete complete sufficient statistic t, thus allowing for an experimental error. Let t have a likelihood

$$p_i(t) = P(t|\lambda^b = \lambda_i^b).$$

The choice of experiment b, the set of possible parameter values $\{\lambda_i^b\}$ and the result t again constitute a question-and answer set, but now in a more advanced form. The point is that the answer is uncertain, so that all these elements together with the likelihood function must be included to specify the question-and-answer.

Proposition 5.5.1. *Exactly this information, the experiment b, the possible answers and the likelihood for the statistic t can be recovered from the effect defined by:*

$$\mathcal{E} = \sum_i p_i(t) v_i^b v_i^{b\dagger}. \tag{5.15}$$

On the other hand, for fixed t every effect $\mathcal{E}$ can be written in the spectral form (5.15).

Proof.

This is a spectral decomposition from which the eigenvalues $p_i(t)$ and the eigenvectors v_i^b can be recovered. As discussed before, the eigenvectors correspond to the question-and-answers for the case without measurement errors, and then the likelihood of the observator t. The last part is obvious.

All this will be further discussed in Section 5.7.

Consider now the situation where a quantum system is known to be in a state given by v_k^a, that is, a perfect experiment a has been performed with result $\lambda^a = \lambda_k^a$. Then make a new experiment b, but let this experiment be non-perfect. We require the probability $\pi(\mathcal{E})$ that the result of the latter experiment shall be t, corresponding to the effect $\mathcal{E}$ given by (5.15). For this situation it is natural to define

$$\pi(\mathcal{E}) = \sum_i p_i(t) P(\lambda^b = \lambda_i^b | \lambda^a = \lambda_k^a). \tag{5.16}$$

An important point in our development is that under Axiom 4, this π, when ranging over all the effects $\mathcal{E}$ will be a generalized probability. The crucial result is the following

Proposition 5.5.2. *Under Axiom 4, if $\mathcal{E}_1$, $\mathcal{E}_2$ and $\mathcal{E}_1 + \mathcal{E}_2$ all are effects, then*

$$\pi(\mathcal{E}_1 + \mathcal{E}_2) = \pi(\mathcal{E}_1) + \pi(\mathcal{E}_2).$$

Proof.
Let $\mathcal{E}_1 = \mathcal{E}$ be given by (5.15), and let

$$\mathcal{E}_2 = \sum_j q_j(t) v_j^c v_j^{c\dagger}$$

for another experiment c with another likelihood q_j.

First we remark that the relations $\pi(r\mathcal{E}_1) = r\pi(\mathcal{E}_1)$ and $\pi(\mathcal{E}_1 + \mathcal{E}_2) = \pi(\mathcal{E}_1) + \pi(\mathcal{E}_2)$ are trivial when $\mathcal{E}_1$, $\mathcal{E}_2$, $r\mathcal{E}_1$ and $\mathcal{E}_1 + \mathcal{E}_2$ are all effects and all $v_i^c = v_i^b$.

We now turn to the general case. The statistic t may then be assumed to be sufficient and complete with respect to both likelihoods. By Axiom 1 the parameters of the two experiments are connected by a group transformation. Then by imitating the argument in the proof of Proposition 3.11.2, a complete sufficient statistic for experiment b can be transformed by an isomorphic group transformation to a complete sufficient statistic for experiment c; hence the complete sufficient statistics for the two experiments may be assumed identical.

Consider the experiment $\mathcal{E}_3$ defined by selecting experiment $\mathcal{E}_1$ with probability $1/2$ and experiment $\mathcal{E}_2$ with probability $1/2$. Since the same measurement apparatus was used in both experiments, one can arrange things in such a way that the person reading t for experiment $\mathcal{E}_3$ does not know which of the experiments $\mathcal{E}_1$ or $\mathcal{E}_2$ is chosen. This arrangement is necessary in order to avoid that the conditionality principle should disturb our argument for this situation; see Aitkin [10] and the response to these comments. We can regard $\mathcal{E}_3$ as a genuinely new experiment here.

Now use Axiom 1. From this Axiom there exists a group element g_{bc} such that $\lambda^c(\phi) = \lambda^b(\phi g_{bc})$. We can, and will, rotate experiment b in such a way that all final state vectors coincides with those of experiment c. Then from Axiom 4, the transition probability to experiment $\mathcal{E}_2$ is the same as if a rotated initial state was chosen, the state vectors v_i^b were chosen, but with a different likelihood $q_i'(t) = q_i(tg_{bc})$.

From this perspective, the experiment $\mathcal{E}_3$ can also be related to the same state vectors, but with a likelihood

$$r_i(t) = \frac{1}{2}(p_i(t) + q_i'(t)). \tag{5.17}$$

The statistic t will be sufficient relative to this likelihood, but may not be complete or minimal. However, this is not needed for our argument.

This gives

$$\pi(\mathcal{E}_3) = \frac{1}{2}\pi(\mathcal{E}_1) + \frac{1}{2}\pi(\mathcal{E}_2) \tag{5.18}$$

for experiments transformed to have the same final states.

We can now transform back so that all three experiments have the same initial state. Since experiment $\mathcal{E}_3$ in the rotated form had the same question-and-answer form as the other two experiments, only with a different likelihood (5.17), this experiment must also correspond to some effect. Then from (5.18), Axiom 4 and the fact that the same sample space is used for all three experiments both in the original and in the rotated version, the transition probability must satisfy

$$\pi(\mathcal{E}_3) = \pi(\frac{1}{2}(\mathcal{E}_1 + \mathcal{E}_2)) = \frac{1}{2}\pi(\mathcal{E}_1) + \frac{1}{2}\pi(\mathcal{E}_2). \tag{5.19}$$

If $\mathcal{E}_1 + \mathcal{E}_2$ is an effect, the factor $1/2$ can be removed throughout by suitably redefining the likelihood.

Proposition 5.5.3. *For fixed initial state $\lambda^a = \lambda_k^a$, the set function defined by (5.16) from the transition probability will under Axiom 4 be a generalized probability on the final effects.*

Proof.
The additivity property for a finite number of effects follows by induction from Proposition 5.5.2. The argument of Proposition 5.5.2 can also be used with a countable set of effects, so the additivity property for generalized effects follows for these set functions.

It is obvious that $\pi(\mathcal{E}) \geq 0$. The limiting effect I corresponds to an experiment and experimental result with likelihood 1 on each single parameter value, and it is clear that the transition probability to this effect must be 1 from every initial state.

<u>Proof of Theorem 5.5.1.</u>

Fix a and k and hence the state v_k^a, interpreted as $\lambda^a = \lambda_k^a$. Define $q_{a,k}(v) = \pi_{a,k}(\mathcal{E})$ to be equal to the transition probability from v_k^a to the effect $\mathcal{E} = vv^\dagger$ for an arbitrary state vector v, assumed to exist in Axiom 4. Generalize to any $\mathcal{E}$ by (5.16). By Proposition 5.5.3 the conditions of the Busch-Gleason theorem are satisfied.

By this theorem, for any $v \in \mathbf{H}$, we have $\pi_{a,k}(vv^\dagger) = v^\dagger \rho v$ for some ρ, which is positive, self-adjoint and has trace 1. This implies $\rho = \sum_j c_j u_j u_j^\dagger$ for some orthogonal set of vectors $\{u_j\}$. Selfadjointness implies that each c_j is realvalued, and positivity demands $c_j \geq 0$ for each j. The trace 1 condition implies $\sum_j c_j = 1$.

Inserting this gives $\pi_{a,k}(vv^\dagger) = \sum_j c_j |v^\dagger u_j|^2$. Specialize now to the particular case given by $v = v_k^a$ for some k. For this case one must have $\sum_j c_j |v_k^{a\dagger} u_j|^2 = 1$, and thus

$$\sum_j c_j (1 - |v_k^{a\dagger} u_j|^2) = 0.$$

This implies for each j that either $c_j = 0$ or $|v_k^{a\dagger} u_j| = 1$. Since the last condition implies $u_j = v_k^a$ (modulo an irrelevant phase factor), and this is a condition which only can be true for one j, it follows that $c_j = 0$ for all other j than the one leading to $u_j = v_k^a$, and $c_j = 1$ for this particular j. Summarising all this, we get $\rho = v_k^a v_k^{a\dagger}$, and Theorem 5.5.1 follows.

The results above are valid and have relevance also outside quantum theory. In Chapter 7, I will sketch a large scale example where, using Born's formula, the prior probability of a second experiment is found, given the result of a first experiment. The main point is that we have an experimental situation with sufficient symmetry across experiments.

By the same proof, Born's formula can be generalised to $P(\mathcal{E}|\lambda^a = \lambda_k^a) = v_k^{a\dagger} \mathcal{E} v_k^a$ for an arbitrary final effect E (cp. also Section 5.6 below). This gives a transition probability from any state vector $v_k^a \in \mathbf{H}$.

Recall that $\mathbf{H}$ was originally defined using perfect experiments. Using Born's formula, it can be seen that a large class of experiments take the same Hilbert space as a point of departure.

5.6   The Electron Spin Revisited

Summary: The theory of this chapter is illustrated in the simplest possible setting: The spin state of an electron, equivalently and more generally the spin state of a spin 1/2 particle, or even more generally, the state of what is called a qubit, which is a central concept for the recent development of quantum informatics.

Recall that in Section 4.7 the electron spin was modelled by an inaccessible three dimensional vector ϕ, and that the observable parts were just $\lambda^a = \text{sign}(\phi \cdot a)$ for given vectors a. For each a this is a quantised parameter taking the value ± 1. The group G acting upon ϕ was the rotation group together with possible changes of scale.

The Hilbert space $\mathbf{L}^a$ connected to a perfect measurement in the direction a is according to Definition 5.2.5 given by all functions of λ^a. This is a two-dimensional vector space, since every such function can be written as

$$I(\lambda^a = -1)f(-1) + I(\lambda^a = 1)f(1).$$

Thus a basis for $\mathbf{H}^a = W\mathbf{L}^a$ can be taken as $(1,0)^T$ and $(0,1)^T$, where the first coordinate is the indicator that λ^a is -1, and the second coordinate is the indicator that λ^a is $+1$.

According to Theorem 5.3.2, a common Hilbert space for all these experiments can be taken to be one fixed $\mathbf{H}$, and by Theorem 5.3.1 there is a representation of the whole group G upon this space. It is well known that the rotation group has a two-valued representation by the group $SU(2)$ of unitary, two-dimensional matrices with determinant one.

This representation can be given explicitly by the Pauli matrices

$$\sigma_x = \begin{pmatrix} 0 & 1 \\ 1 & 0 \end{pmatrix}, \quad \sigma_y = \begin{pmatrix} 0 & -i \\ i & 0 \end{pmatrix}, \quad \sigma_z = \begin{pmatrix} 1 & 0 \\ 0 & -1 \end{pmatrix}. \tag{5.20}$$

For instance, a rotation transforming $\phi_0 = (1,0,0)^T$ to $\phi = (\alpha, \beta, \gamma)^T$ in the plane spanned by these two vectors, can be characterised by the two-dimensional matrix

$$\begin{pmatrix} \gamma & \alpha - 1 - i\beta \\ \alpha - 1 + i\beta & -\gamma \end{pmatrix} = (\alpha - 1)\sigma_x + \beta\sigma_y + \gamma\sigma_z. \tag{5.21}$$

By transforming by these unitary matrices, we get the important result:

Proposition 5.6.1.
All unit vectors in the electron spin Hilbert space $\mathbf{H}$ are possible state vectors.

Each of these state vectors gives a crisp answer to a question of the type: What is the value of λ^b? The answer being of the form $\lambda^b = -1$ or $\lambda^b = +1$.

In a more common quantum-mechanical language this means that the operator $T^b = b \cdot \sigma = b_x\sigma_x + b_y\sigma_y + b_z\sigma_z$ for a real-valued vector $b = (b_x, b_y, b_z)$ has eigenvalues ± 1, and the eigenvectors have a state vector interpretation corresponding to a perfect spin measurement in the direction b. Let these two eigenvectors, which will be orthogonal, be v_1^b and v_2^b.

This means that the two one-dimensional projections $v_1^b v_1^{b\dagger}$ and $v_2^b v_2^{b\dagger}$ will be orthogonal. Many textbooks discuss the Bloch sphere representations of these projections:

$$v_1^b v_1^{b\dagger} = \frac{1}{2}(I + b \cdot \sigma), \tag{5.22}$$

$$v_2^b v_2^{b\dagger} = \frac{1}{2}(I - b \cdot \sigma). \tag{5.23}$$

The transition probabilities between states defined by spin in different directions are found from the Born formula, which gives:

Proposition 5.6.2

$$P(\lambda^b = +1|\lambda^a = +1) = \frac{1}{2}(1 + \cos(a,b)). \qquad (5.24)$$

Proof.

$$|v_1^{a\dagger} v_1^b|^2 = \text{tr}(v_1^a v_1^{a\dagger} v_1^b v_1^{b\dagger}) = \text{tr}(\frac{1}{4}(I + a \cdot \sigma)(I + b \cdot \sigma))$$

$$= \text{tr}(\frac{1}{4}(I + (a + b) \cdot \sigma + a \cdot bI)) = \frac{1}{2}(1 + a \cdot b).$$

The state as a question-and-answer pair can in fact be represented in many ways. One way is simply by the 3-vector $u = \lambda^a a$. We can recover both the question and the answer from this vector, since a spin component -1 in the direction a is equivalent to a spin component $+1$ in the direction $-a$.

Finally then, it is obvious that a specification of u is equivalent to a specification of the Bloch sphere matrix

$$\rho = \frac{1}{2}(I + u \cdot \sigma),$$

which again by (5.22) and (5.23) is equivalent to specifying the state vector. Thus, summarising:

Proposition 5.6.3.
The spin state can be given in any of four different ways: (1) as a question a together with an answer λ^a; (2) by the 3-vector u; (3) by the Bloch sphere matrix ρ; (4) by the Hilbert space state vector v.

5.7 Statistical Inference in a Quantum Setting

Summary: In this section the rules for statistical inference in a quantum-mechanical setting are derived from the results of this chapter and discussed from several points of view. The concepts of density matrix and effect are connected to experiments. Hypothesis testing with a single hypothesis and a single alternative is generalised to the electron spin setting. Operator valued measures are introduced and showed to have a natural interpretation.

I have stated repeatedly that a state vector v_k^a can be interpreted as a focused question: What is λ^a? together with a crisp answer: $\lambda^a = \lambda_k^a$. As k varies, the vectors v_k^a form an orthonormal basis of the Hilbert space $\mathbf{H}$. Since the phase of v_k^a is irrelevant, one might as well say in general that the question-and-answer pair is equivalent to the one-dimensional projection

$$\rho_k^a = v_k^a v_k^{a\dagger},\tag{5.25}$$

where the last vector may be thought of as v_k^a, transposed and complex conjugated.

Now I want to go one step further: In many cases one can pose a question, and the answer is uncertain in the sense that we only have a probability distribution over the possible answers. In the experimental setting that we are discussing here, we can imagine 3 possible such situations: 1) We have a prior probability distribution π_k over the possible answers; 2) After the experiment we have a posterior probability distribution π_k over the possible answers; 3) Another experiment b has been performed, and from the result of this, the probability distribution of the result of our experiment is given by Born's formula

$$\pi_i = P(\lambda^b = \lambda_i^b | \lambda^a = \lambda_k^a) = |v_k^{a\dagger} v_i^b|^2.$$

In all these cases we can, in the manner common in quantum physics, introduce the uncertain state by the density matrix.

Definition 5.7.1. *The* density matrix *is given by*

$$\rho = \sum_k \pi_k v_k^a v_k^{a\dagger}.\tag{5.26}$$

From this formula one can infer the choice of perfect experiment a, the possible parameter values and the probability of each parameter value. On the other hand, every positive selfadjoint operator with trace (sum of eigenvalues) equal to 1 is a possible density matrix. Note that the crisp states defined by (5.25) are special cases.

The next step is to turn to real, non-perfect experiments. Then, given the choice of experiment a and the parameter value λ_k^a, we assume an ordinary statistical model. Since the parameter is discrete, we limit ourselves here to the case with discrete data, also. Assume that the probability of y^a, given the experiment a with parameter value λ_k^a, is $p(y^a | \lambda_k^a)$.

The definition implied by Proposition 5.5.1 is repeated here for convenience.

Definition 5.7.2. *The effects are defined as the operators*

$$\mathcal{E}(y^a) = \sum_k p(y^a | \lambda_k^a) v_k^a v_k^{a\dagger}.\tag{5.27}$$

For each given data point y^a the effect $\mathcal{E}(y^a)$ is a selfadjoint positive operator. Note that this is the spectral decomposition of the operator $\mathcal{E}(y^a)$, so, given this

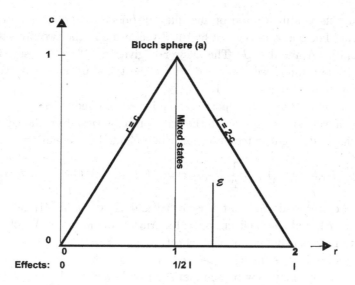

Fig. 5.1   An effect $\mathcal{E}$ in the (r, c)-plane.

operator, we can recover the eigenvalues $p(y^a|\lambda_k^a)$ and the eigenvectors v_k^a. Thus we can read from the operators which experiment, that is, question and possible answers, is performed, together with the statistical model for given data, in statistical terms the likelihood. In other words, the operator $\mathcal{E}(y^a)$ contains all the information that is necessary for doing statistical inference.

Let us first specialise this to the case of a parameter taking only two possible values, like in the electron spin case. We let the two values be $+1$ and -1. From a statistical point of view this also corresponds to the Neyman-Pearson case where one can test a single hypothesis against a single alternative. So we will look upon a hypothesis testing problem in this situation with level α and power β.

In our connection this means the following: First the experiment a is chosen. Then, before any data are obtained, we make a programme stating how our decision procedure shall be. This goes as follows:

1) The decision shall be based upon an observator t^a, a function of the observations, which also takes the values $+1$ or -1,

2) This choice of t^a shall be made in such a way that the two error probabilities are fixed: If the correct parameter is λ^a, then $P(t^a = -1|\lambda^a = +1) = \alpha$ and $P(t^a = +1|\lambda^a = -1) = 1 - \beta$.

In common statistical language this means that we are testing the hypothesis $H_0 : \lambda^a = +1$, and this hypothesis is rejected if $t^a = -1$. Then α is the *level* of the test, the probability of wrong rejection, while β is the *power* of the test, the probability of rejecting the hypothesis when we should.

Note that this is still a state of the question/answer type, albeit in a more advanced form: The question is given by the 3-vector a and the two predetermined error probabilities α and $1 - \beta$. The answer is given by $\lambda^a = +1$, say, which is the conclusion we claim if we observe $t^a = +1$. So the state must involve all the quantities b, α, β and the answer ± 1.

Say that we have done the experiment and reported the value $+1$. Then we again will use a weighing according to the error probabilities, even though these at the outset refer to different outcomes. Thus the weighted state will be

$$\mathcal{E} = (1-\alpha)\frac{1}{2}(I + b \cdot \sigma) + (1 - \beta)\frac{1}{2}(I - b \cdot \sigma) = \frac{1}{2}((2 - \alpha - \beta)I + (\beta - \alpha)b \cdot \sigma). \quad (5.28)$$

This state corresponds to what Busch [38] and Caves et al [41] call an effect $\mathcal{E} = \frac{1}{2}(rI + cb \cdot \sigma)$, and these effects played a crucial rôle in our proof of the Born formula. In terms of the definition of an effect, we have $r = 2 - \alpha - \beta$ and $c = \beta - \alpha$. The hypothesis problem is further discussed in Section 7.6.

Our state concept may now be summarised as follows: To the state $\lambda^a(\cdot) = \lambda_k$ there corresponds the state vector v_k^a, and these vectors determine the transition probabilities by Born's formula. This formula also implies for perfect experiments:

Theorem 5.7.1. (a) $E(\hat{\lambda}^b | \lambda^a = \lambda_k^a) = v_k^{a\dagger} T^b v_k^a$, where $T^b = \sum \lambda_j^b v_j^b v_j^{b\dagger}$.
(b) $E(f(\hat{\lambda}^b)|\lambda^a = \lambda_k^a) = v_k^{a\dagger} f(T^b) v_k^a$, where $f(T^b) = \sum f(\lambda_j^b) v_j^b v_j^{b\dagger}$.

Proof.

$$E(f(\hat{\lambda}^b)|\lambda^a = \lambda_k^a) = \sum_i f(\lambda_i^b) P(\lambda^b = \lambda_i^b | \lambda^a = \lambda_k^a)$$

$$= \sum_i f(\lambda_i^b)|v_i^{b\dagger} v_k^a|^2 = \sum_i f(\lambda_i^b) v_k^{a\dagger} v_i^b v_i^{b\dagger} v_k^a.$$

Thus, in ordinary quantum-mechanical terms, the expectation of every observable in any state is given by the familiar formula.

Now turn to non-perfect experiments. In ordinary statistics, an experiment is specified by a probability measure $P^\theta(dy)$ depending upon a parameter θ. Assume now that such a measurement depends upon the parameter λ^b, while the current state is given by $\lambda^a = \lambda_k^a$. Then as in Theorem 5.7.1 (b):

Theorem 5.7.2. (a) *Corresponding to the experiment $b \in \mathcal{A}$ one can define an operator-valued measure $\mathcal{M}$ by $\mathcal{M}(dy) = \sum_j P^{\lambda_j^b}(dy) v_j^b v_j^{b\dagger}$. Then, given the initial state $\lambda^a = \lambda_k^a$, the probability distribution of the result of experiment b is given by $P[dy|\lambda^a = \lambda_k^a] = v_k^{a\dagger} \mathcal{M}(dy) v_k^a$.*

(b) These operators satisfy $\mathcal{M}[S] = I$ *for the whole sample space* S, *and furthermore* $\sum \mathcal{M}(A_i) = \mathcal{M}(A)$ *for any finite or countable sequence of disjoint elements* $\{A_1, A_2, \ldots\}$ *with* $A = \cup_i A_i$.

Theorem 5.7.2 (b) is easily checked directly. The proof of Theorem 5.7.2 (a) is similar to that of Theorem 5.7.1.

A more general state assumption is a Bayesian one corresponding to this setting. From Theorem 5.7.2 (a) we easily find:

Theorem 5.7.3. *Let the current state be given by probabilities* $\pi(\lambda_k^a)$ *for different values of* λ_k^a. *Then, defining* $\rho = \sum \pi(\lambda_k^a) v_k^a v_k^{a\dagger}$, *we get* $\mathrm{P}[dy] = \mathrm{tr}[\rho \mathcal{M}(dy)]$.

A density operator ρ of such a kind is often used in quantum mechanics; the definition above gives a precise interpretation. In fact, these results are the basis for much of quantum theory, in particular for the quantum statistical inference in Barndorff-Nielsen et al.[18]; for a formulation, see also Isham [130].

Note that the density matrix $v_k^a v_k^{a\dagger}$ is equivalent to the pure state v_k^a; similarly, a density matrix $v_j^b v_j^{b\dagger}$ is equivalent to the statement that a perfect measurement giving $\lambda^b = \lambda_j$ just has been performed. By straightforward application of Born's formula one gets:

Theorem 5.7.4. *a) Assume an initial state* v_k^a, *and assume that a perfect measurement of* λ^b *has been performed without knowing that value. Then this state is described by a density matrix* $\sum_j |v_k^{a\dagger} v_j^b|^2 v_j^b v_j^{b\dagger}$.

b) After measurement $\lambda^b = \lambda_j^b$ *the state vector then changes to* v_j^b. *From a) this happens with probability* $|v_k^{a\dagger} v_j^b|^2$.

This is related to the celebrated and much discussed projection postulate of von Neumann.

In general we have assumed for simplicity in this section that the state vectors are non-degenerate eigenvectors of the corresponding operators, meaning that the parameter λ^a contains all relevant information about the system. This can be generalised, however.

Axiom 5. *Whenever a measurement has resulted in a value* λ_k *for some parameter* λ *or in some measurent data related to this, all subsequent measurements will be consistent with these observations.*

This is in agreement with the context concept, which will be further discussed in Section 5.10 below.

5.8   Proof of the Quantum Rules from Our Axioms

Summary: *Concluding one of the main goals of this book, I show that the first 3 basic Quantum rules as formulated in [130] follow from the results found in this chapter, and hence from our axioms. The 4th Quantum rule will be adressed in the next chapter.*

For convenience I repeat the three first quantum rules from Chapter 4 as they are also given by Isham [130]:

Quantum rule 1. *The predictions of results of measurements made on an otherwise isolated system are probabilistic in nature. In situations where the maximum amount of information is available, this probabilistic information is represented mathematically by a vector in a complex Hilbert space* **H** *that forms the* state space *of the quantum theory. In so far as it gives the most precise predictions that are possible, this vector is to be thought of as the mathematical representation of the physical notion of 'state' of the system.*

Quantum rule 2. *The observables of the system are represented mathematically by self-adjoint operators that act on the Hilbert space* **H**.

Quantum rule 3. *If an observable quantity $\hat{\lambda}^a$ is represented by the self-adjoint operator T^a, and the state by the normalised vector $v \in$* **H**, *then the expected result of the measurement is*

$$E_v(\hat{\lambda}^a) = v^\dagger T^a v. \tag{5.29}$$

The following result is quite simple now, but crucial for the purpose of the entire book:

Theorem 5.8.1. *In the c-system given by Definition 4.6.1, the Quantum rules 1-3 follow from our Axioms 1-4.*

Proof.
The state vector of Quantum rule 1 is defined in the simplest case in Definition 5.4.1 and in general in Definition 5.4.2. The sense in which we interpret this vector to correspond to a state of the given system, is connected to a question-and-answer pair: A maximal question of the form: What is the value of λ^a? is asked, and a crisp answer: $\lambda^a = \lambda_k^a$ is given. By Theorem 5.4.2, the state vector corresponds in a unique way to such a question-and-answer pair.

Not all unit vectors are possible state vectors. Superselection rules are discussed at the end of Section 5.4.

By Quantum rule 2, the operator characterising an observable (in our conceptual framework a parameter) is defined in equation (5.12), and its basic properties are given in Theorem 5.4.1.

In principle the probability distribution over a parameter ('observable') may be of one of three kinds: 1) A prior probability; in our setting usually the invariant measure corresponding to the basic symmetry group; 2) a posterior distribution after the observation of data; of 3) a transition probability found from Born's formula. In the case where we start with a state vector v, the last possibility is the actual one, and the expectation result of Quantum rule 3 is given by Theorem 5.7.1. The cases 1) and 2) concern a single parameter, and the expectation may then also be written in the same way, but reduces to a simpler form.

There are variants of the basic quantum rules in the literature; see for instance Holevo [126], but apart from a possible adjustment of technicalities, they can all be deduced in a similar manner from our axioms.

5.9 The Case of Continuous Parameters

Summary: An informal discussion of the case of one or two continuous parameters are given and illustrated on the case of position and momentum for a single, one-dimensional particle. The example of the two-slit experiment is discussed from this point of view.

Some of the results of the previous sections are true immediately also for parameters varying over a continuous range, for instance the whole real line. It was not until Section 5.4 when the state vectors should be constructed and given an interpretation, that Axiom 3 on discreteness was needed. To handle continuous parameters rigorously at this point, we need more advanced mathematical constructions like rigged Hilbert spaces, von Neumann algebras or C^*-algebras.

In this section I will instead give a non-rigorous discussion using slightly informal limiting operations and pretending that delta-functions and complex exponentials can be taken as belonging to a Hilbert space of square-integrable functions.

Start with a discrete parameter ξ^a taking values on the grid $\{0, \pm a, \pm 2a, ...\}$, and associate with this parameter the translation group G^a with elements g_j given by $ka \rightarrow (k+j)a$ for $k, j = 0, \pm 1, \pm 2, ...$ By analogy with Definition 5.4.1 the statements $\xi^a = ka$ can be associated with the rather trivial state function

$$f^a(\xi) = 1 \text{ when } \xi = ka, \tag{5.30}$$

and $f^a(\xi) = 0$ elsewhere, that is, for $\xi = ia, i \neq k$.

I could let $a \rightarrow 0$ at this point, but that would leave us with a rather trivial system. Hence I will first add to the system a complementary parameter π^b, also taking values in some grid $\{0, \pm b, \pm 2b, ...\}$, and let $\phi^a b$ be the c-variable (ξ^a, π^b). I assume that $\phi^a b$ is inaccessible, but that each of the parameters ξ^a and π^b are accessible.

Now let $a, b \to 0$ resulting in continuous parameters ξ and π, which can be interpreted as the position and momentum of a one-dimensional particle. From results proved in Chapter 6 one can show:

a) There is a non-commutative group G acting upon the c-variable $\phi = (\xi, \pi)$ so that the subgroup acting upon ξ is given by the translations $\xi \to \xi + c$.

b) In the space $\mathbf{H}$ of functions of ξ the state functions of this parameter are delta functions, and the operator X corresponding to this parameter is a multiplication by ξ.

c) The generator on the same space of the subgroup acting upon π is given by the differential operator

$$P = \frac{\hbar}{i} \frac{\partial}{\partial \xi},$$

where $\hbar$ is a universal constant (Planck's constant; see Proposition 6.5.3). This is also the operator corresponding to the parameter π. The state function corresponding to $\pi = \pi_0$ is

$$v(\xi) = \exp(\frac{i\pi_0 \xi}{\hbar}). \qquad (5.31)$$

d) This is a standing wave with wavelength $\lambda = \frac{\hbar}{\pi_0}$, so a state with constant momentum π_0 can be associated with such a wave.

e) It is straightforward to show that

$$PX - XP = \frac{\hbar}{i},$$

from which it follows easily that the Heisenberg inequality of Theorem 1.1.1 holds for a perfect measurement of position ξ and momentum $\pi = mv$. If measurement error is included in these measurements, assuming for simplicity unbiased measurements x and p, we have

$$Var(x) = Var(E(x|\xi)) + E(Var(x|\xi)) \geq Var(\xi),$$

$$Var(p) = Var(E(p|\pi)) + E(Var(p|\pi)) \geq Var(\pi).$$

So Heisenberg's inequality also holds for the real measurements.

Example 5.9.1. The two-slit experiment: I will use the results above to give a somewhat simplified discussion of an example found in many textbooks.

Assume that a single particle is sent towards a screen 1 with two small, parallel slits. Behind this screen there is another one, screen 2, so that the position of the particle there is registered. This experiment is repeated many times, so that one gets a pattern on screen 2. When both slits are open, we get a wave pattern.

Now the point is: In whatever way we try to determine which slit the particle went through, the wave pattern disappears, and we get a smooth distribution on screen 2.

Let us concentrate on the coordinate ξ of a single particle just outside screen 1, parallel to the screen and perpendicular to the slits. Let π be the component of momentum of the particle in the same direction. Apart from the fact that ξ now has been limited to one of the coordinates ξ_1 or ξ_2 of the slits, these quantities could as well have been measured just inside the screen.

If both slits are open, we have no certain information about ξ. The Heisenberg inequality will indicate that one cannot exclude the possibility of some information about π. A definite, certain information about π will amount to a standing wave, but a more limited information will also result in a wave pattern in this direction, which is transferred to screen 2.

Now let us assume that we try to get some information about which slit the particle goes through. Then we ask the question: What is the value of ξ? - and we allow ourselves to get a definite answer: ξ_1 or ξ_2. In that case the variance of ξ will be zero, and Heisenberg's inequality forces the variance of π to be infinite. Hence a wave pattern then is impossible.

One reason why this example has caused so much discussion in the literature, is that ξ and π are interpreted as some real quantities associated with a particle which is attempted to be visualized in some classical mechanical sense. Imagining such quantities as parameters for potential measurements, i.e., c-variables, is more abstract, but it facilitates interpretation.

5.10   On the Context of a System, and on the Measurement Process

Summary: Every state is related to a context: The physical environment, the result of previous measurements, the concepts understood when posing the question determining the state and so on. A brief discussion of the context concept is given. Then I develop in detail an alternative approach to the quantum Hilbert space based upon the sample space available for the measurement. Finally the measurement process itself is briefly discussed using previous results.

The first step towards the determination of a state is to focus on a question to ask about the system under consideration. As I have formulated it, the question is about the value of some statistical parameter, and it shall be a maximal such parameter, that is, contain as much information as possible. In the discussion which follows, it may be enlightening to exchange the word 'parameter' with 'concept', as given by a set of words that can be used to describe the system.

Now it is crucial that the words we use to describe some system, depend on our previous knowledge, in short, the properties about the system that we consider to be obvious. In a quantum physics setting, these can be absolutely conserved quantum numbers such as mass or charge, or they can be inferred from previous measurements. Also of relevance may be intrinsic limitations on what can be measured in the system, or such limitations imposed by the environment. It is then assumed that these limitations are absolute, that is, they cannot be changed by refinements of the measurement procedure.

All these factors constitute what will be called the context of the system. For the validity of the quantum rules, as we see it here, it is necessary that our 5 axioms are valid *in the relevant context*.

One contextual limitation may be caused by the very possibility of measurement. Such a limitation may be due to the measurement apparatus, which is less interesting in our setting, but it may also be due to intrinsic properties of the system itself. I will indicate by a series of mathematical results that such limitations may give one way leading to the quantum-mechanical Hilbert space.

Concentrate on a single measurement y with a statistical model P^λ, whose parameter depends upon a c-variable: $\lambda = \lambda(\phi)$. Call the sample space S, and let the model be dominated by some measure P.

We use the ordinary concept of sufficiency, see, e.g., Lehmann and Casella [146], also discussed in Section 2.11 above; the definition is repeated here for convenience:

Definition 5.10.1. *A random variable $t = t(\omega); \omega \in S$ connected to a model P^λ is called sufficient if the conditional distribution of each other variable y, given t, is independent of the parameter λ.*

A sufficient statistic t is minimal if all other sufficient statistics are functions of t. It is complete if

$$E^\lambda(h(t)) = 0 \text{ for all } \lambda \text{ implies } h(t) \equiv 0 (\text{for all} h). \qquad (5.32)$$

It is well known that a minimal sufficient statistic always exists and is unique except for invertible transformations, and that every complete sufficient statistic is minimal. If the statistical model has a density belonging to an exponential class

$$b(y)d(\lambda)e^{c(\lambda)'t(y)},$$

and if $c(\Lambda) = \{c(\lambda) : \lambda \in \Lambda\}$ contains some open set, then the statistic t is complete sufficient.

Recall that a function $\xi(\lambda)$ is called unbiasedly estimable if $E^\lambda(y) = \xi(\lambda)$ for some y. Given a complete sufficient statistic t, every unbiasedly estimable function $\xi(\lambda)$ has one and only one unbiased estimator that is a function of t. This is the unique unbiased estimator with minimum risk under weak conditions [146]. Thus complete sufficiency leads to efficient estimation.

Assumption 5.10.1. *The experiment can be chosen in such a way that there is a complete sufficient statistic t under the model P^λ.*

In the following I write D for G^a, which was the subgroup corresponding to the selection of the fixed experiment a. I keep a fixed here. D will be a fixed group on the common sample space S, but also acts on the selected parameter space.

Definition 5.10.2. *The Hilbert space* **K** *is defined as the set of all functions* $h(t)$ *such that* $h(t) \in L^2(S, P)$ *and* $f(\phi) = E^{\lambda(\phi)}(h(t)) \in L^2(\Phi, \nu)$.

In this definition the function h is assumed to be complex-valued. It is easy to see that equation (5.32) holds for complex functions if and only if it holds for realvalued functions.

A sufficient condition for $f \in L^2(\Phi, \nu)$ is that $\int E^{\lambda(\phi)}(|h(t)|^2)\nu(d\phi) < \infty$. Since it is defined as a closed subspace of a Hilbert space, the Hilbert space property of **K** is seen to hold.

From a statistical point of view it is very satisfactory that the sufficient statistic determines the Hilbert space for single experiments. The sufficiency principle, by many considered to be one of the backbones of statistical inference (e.g. [25]) says that identical conclusions should be drawn from all sets of observations with the same sufficient statistic. It is also of importance that this Hilbert space satisfies the invariance properties that are needed in order that it can serve as a representation space for the symmetry groups connected to each experiment.

Let then the group D be acting upon the sample space S, on the parameter space Λ and on the c-variable space Φ. Recall the brief discussion of group representations in Chapter 3. In particular, recall the definition of the space V_λ, an invariant space under the regular representation of the group D on $L^2(\Phi, \nu)$.

Proposition 5.10.1. *The space* **K** *is an invariant space for the regular representation of the observational group* D *on* $L^2(S, P)$, *i.e., under* $U(g)h(t) = h(tg)$; $g \in D$.

Consider now the operator A from **K** to $V_\lambda \subset L^2(\Phi, \nu)$ defined by

$$(Ay)(\lambda(\phi)) = \int y(\omega)P^{\lambda(\phi)}(d\omega) = E^{\lambda(\phi)}(y), \tag{5.33}$$

using again the (reduced) model $P^\lambda(d\omega)$ corresponding to the experiment a. In the following it will be important to use **K** to construct a Hilbert space related to the parameter space.

Definition 5.10.3. *Define the space* **L** *by* **L** $= A$**K**.

By the definition of a complete sufficient statistic, the operator A will have a trivial kernel as a mapping from **K** onto A**K**. Hence this mapping is one-to-one. It is also continuous and has a continuous inverse. (See below.) Hence **L** is a closed subspace of $L^2(\Phi, \nu)$, and therefore a Hilbert space. Note also that **L** is the space in $L^2(\Phi, \nu)$ of unbiasedly estimable functions with estimators in $L^2(S, P)$. It is in general included in the space V_λ of all functions of the parameter λ.

Proposition 5.10.2. *The space* **L** *is an invariant subspace of* $L^2(\Phi, \nu)$ *for the regular representation of the group* D *on* $L^2(\Phi, \nu)$.

A main result proved in Appendix A.3.2 is now:

Theorem 5.10.1. *The spaces* $\mathbf{K} \subset L^2(S, P)$ *and* $\mathbf{L} \subset L^2(\Phi, \nu)$ *are unitarily related. Also, the regular representations of the group D properly defined on these spaces are unitarily related.*

This means that either $\mathbf{K}$ or $\mathbf{L}$ can be taken as a basic Hilbert space connected to the experiment. Since we started by constructing $\mathbf{K}$, this approach definitely takes the sample space and the possible limitations due to the measurement as a point of departure.

Now return to the situation where one selects an experiment a among a class of experiments $\mathcal{A}$. Corresponding to this choice we have a parametric Hilbert space $\mathbf{L}^a$ and an observational Hilbert space $\mathbf{K}^a$ with a unitary relation between them. One interesting thing is that for a perfect measurement, that is, one where measurement error can be neglected, we come back to the basic Hilbert space of Theorem 5.3.1, which was the main component of the quantum-mechanical Hilbert space as constructed above.

Proposition 5.10.3. *For a perfect experiment the space $\mathbf{L}^a$ is just the space of functions $\tilde{f}$ of $\lambda^a(\cdot)$ such that $f(\phi) = \tilde{f}(\lambda^a(\phi)) \in L^2(\Phi, \nu)$.*

In general, the common Hilbert space is constructed as in Section 5.3, and the discussion of the following sections is valid in this context, also.

Other approches to the context concept and to the measurement process can be inferred from the quantum-mechanical literature. Much of this can be related to our axioms and results. As already remarked, Axiom 5 in Section 5.7 is closely related to our discussion of context.

The well known measurement formula

$$P[dy] = \text{tr}[\rho \mathcal{M}(dy)]$$

is proved in Theorem 5.7.3 using a natural definition of the operator valued measure $\mathcal{M}$. To say something about the state after measurement, it is common in quantum theory to introduce the abstract notion of an *instrument*; see the discussion in Barndorff-Nielsen et al. [18]. A very simple beginning of this theory is given in Theorem 5.7.4.

Chapter 6

FURTHER DEVELOPMENT OF QUANTUM MECHANICS

6.1 Introduction

From Chapter 5 we now have a mathematical correspondence between the conventional quantum theory formalism and our own formalism based upon questions, answers and symmetries. The construction of a Hilbert space for a single experiment in Section 5.2 was simple, but to glue these together, I needed the symmetry assumption in Axiom 1. This then resulted in a common Hilbert space which also was an invariant space for the basic group G. The unit vectors of this space could be taken as state vectors in the usual way, and they could be uniquely interpreted as maximal questions plus answers. Exceptions to this were given by superselection rules. Operators corresponding to parameters (observators) were constructed in the ordinary way, and had the ordinary quantummechanical interpretation. Probabilities were found from the Born formula, which again was proved using symmetry assumptions. For related derivations, partly from other assumptions see references in Section 5.5. One advantage of my proof is nevertheless that it goes in a natural way via the *effects*, which have a place in quantummechanical inference. A starting point for this inference is provided in Section 5.7. The last sections of Chapter 5 give some further developments, and in this Chapter I will continue in different directions along the same route.

In discussions about the interpretations of quantum mechanics and in other philosophical discussions one often distinguishes between the epistimological - what is knowledge-dependent - and the ontological - what really *is*. The development above leads to an interpretation of quantum theory which may have some elements of ontology, but is far from fully ontological at the outset: We do not have any ambition to infer from a given state in detail how nature *is*, nor is such a detailed description possible. The states are connected to focused questions about nature, hence to our choices, and the answers are therefore connected to our perception of nature. This leads in principle to an epistemological interpretation. Nevertheless, if a maximal question can be answered through a *perfect* experiment, we have perfect information about the relevant parameter, hence perfect partial information about some given system. In the ordinary quantummechanical way, detailed information about other parameters is then precluded.

The very fact that we from Chapter 5 have two equivalent formalisms, the ordinary Hilbert space formalism and the present question-and-answer formalism, shows that quantum theory can be described in several different languages. This in itself indicates that an epistemological interpretation is in place. In addition, as just explained, my own formulation in an obvious way points in the same direction.

In the present Chapter I will take up various further aspects of quantum mechanics, in order to complete the theory: Entanglement, Bell's inequality, Planck's constant, the Schrödinger equation, quantum histories, relation to the many-world interpretation and to some so-called paradoxes. Of course, many more topics could have been taken up, but I feel that the issues treated here are sufficient to prove my point. My aim here is essentially to show that the approach of this book is capable of leading to a consistent, complete theory. The approach is new, however, and since conventional quantum theory is much more developed, I will rely upon this theory at various technical points.

6.2   Entanglement

Assume that we are interested in the combined state of two systems, say two particles as described by their spin states. As isolated systems the two systems are assumed to be in some states v_k^a and u_j^b, respectively, the first one indicating that the question about some parameter λ^a has been given the answer λ_k^a, and the second one characterising some $\mu^b = \mu_j^b$. In the simplest case then the combined state is $v = v_k^a \otimes u_j^b$. Since we can consider this extended physical object as a unit, this state gives answers to both questions: What is λ^a? and What is μ^b?

Now the Hilbert space spanned by these product state vectors contains all linear combinations of them. As we saw in Chapter 5, at least if we disregard superselection rules and similar phenomena, all unit vectors of the Hilbert space are possible state vectors, and so in particular in the present case are the normalised linear combinations of the product states. Any such linear combination which is not a simple product state, is called an *entangled* state. Physically, entangled states can develop from product states through the Schrödinger equation, which will be discussed later in this chapter. At this point we will just say that the Schrödinger equation is linear, so linear combination of product states evolves into linear combination of product states. Entangled states may also in principle be interpreted as the answers of some questions; sometimes they are eigenstates of simple variables like total spins.

Consider now the case where λ^a and μ^b are spin parameters of two qubits, i.e., spin 1/2 particles like the electron, and consider the entangled state where the total spin is zero. It turns out that such a state exists even though we do not know a priori that the total spin vector is an accessible parameter. Indeed we might characterise this state by saying that the spin component of the combined particle in any direction is zero, and such spin components are individually known to be accessible. A more concise argument can be given within conventional quantum theory.

Then we have a rather obvious phenomenon which nevertheless has some strange implications: If λ^a is measured in a perfect measurement to be $+1$, then by necessity, if we make a perfect measurement of μ^b in the direction $b = a$, the answer must be -1. And oppositely, $\lambda^a = -1$ implies $\mu^a = +1$.

Note that the determination of the total spin as zero could have been done in some distant past when the two particles where close together. But when the measurements above were done, the particles could in principle be at a large distance from each other. The situations become really strange when one considers that different directions a can be chosen for the spin component measurement at the first particle, and that this is a free choice by an observer associated with this particle. Then we apparently have transference of information over a distance. This leads to the essence of what is called the Einstein, Podolsky, Rosen - paradox [74].

Thus observations on entangled states may naively seem to be in conflict with Einstein's relativity theory, which says that information cannot be transferred faster than the speed of light. It can be proved in standard quantum theory, however, that although two entangled systems apparently may send information information across large spatial distances, no useful information can be transmitted in this way, so causality cannot be violated through entanglement. This is the content of the recent no communication theorem [173].

Entangled states are the building blocks of most of what is done in the vast and rapidly developing field of quantum informatics, an area which will not be discussed in this book. One of the themes of quantum informatics is quantum computers, computers that in theory shall be very efficient, but where much remains to be done in practice.

6.3 The Bell Inequality Issue

Consider then the situation of Einstein, Podolsky and Rosen [74] as modified by Bohm, where two particles previously have been together in a spin 0 (singlet) state, so that they - in our notation - later have opposite spin vectors modeled by the inaccessible vector c-variables ϕ and $-\phi$. Explicitly, the state can be written as

$$w = \frac{1}{\sqrt{2}}(u_+v_- - u_-v_+),$$

where u_+v_- is a state where particle 1 has spin component $+1$ and particle 2 has spin component -1 along the z-axis, and *vice versa* for u_-v_+. As just discussed, this is described as an entangled state, that is, a state for two systems which is not a direct product of the component state vectors. I will here follow my own programme, however, and stick to the parametric description.

As pointed out by Bell [23] and others, correlation between distant measurements may in principle be attributed to common history, but this can not be the full explanation in this case, where Bell's inequality may be violated.

Assume that spin components λ^a and μ^b are measured in the directions given by the unit vectors a and b on the two particles at distant sites A and B, where the measured values $\hat\lambda^a$ and $\hat\mu^b$ each take values ± 1. Let this be repeated 4 times: Two settings a, a' at site A are combined with two settings b, b' at site B. The socalled CHSH version of Bell's inequality then reads:

$$\mathrm{E}(\hat\lambda^a\hat\mu^b) \leq \mathrm{E}(\hat\lambda^a\hat\mu^{b'}) + \mathrm{E}(\hat\lambda^{a'}\hat\mu^b) + \mathrm{E}(\hat\lambda^{a'}\hat\mu^{b'}) + 2. \tag{6.1}$$

In fact we can easily show the seemingly stronger statement:

$$\hat\lambda^a\hat\mu^{b'} + \hat\lambda^{a'}\hat\mu^b + \hat\lambda^{a'}\hat\mu^{b'} - \hat\lambda^a\hat\mu^b = \pm 2 \tag{6.2}$$

whenever all estimates take the values ± 1:

Lemma 6.3.1. *The equality (6.2) holds in the setting sketched above.*

Proof.

All the products take values ± 1 and $\hat\lambda^a\hat\mu^b$ is the same as the product of the first three products. Listing the possibilities of signs here, then shows that the left-hand side of (6.2) always equals ± 2.

As is well known, the inequality (6.1) can be violated in the quantum-mechanical case, and this is also well documented experimentally. There is a large literature on Bell's inequality, and I will not try to summarise it here. The derivation of (6.2) above seems quite obvious, and the usual statement in the quantum-mechanical literature is that (6.1) follows under what is called local realism. Because of the experimental evidence against (6.1), most physicists are of the opinion that quantum mechanics does not satisfy local realism. But there are also some [93, 142] who claim that there may be loopholes that remain to be closed in the experiments, and that they therefore will not support the conclusion that (6.1) is definitively shown to be broken.

The following is an important part of my own philosophy: Quantum theory is a statistical theory, and should be interpreted as such. In that sense the comparison to a classical mechanical world picture, and the term 'local realism' inherited from this comparison is not necessarily of interest. I am more interested in the comparison of ordinary statistical theory and quantum theory. My aim is that in it principle should be possible to describe both by essentially similar ways of modelling and inference. Thus it is crucial for me to comment on the transition from (6.2) to (6.1) from this point of view.

As pointed out by Gill [94], for any way that the experiment is modelled by replacing the spin measurements by random variables, there is no doubt that this transition is valid, and the inequality (6.1) must necessarily hold. The reason is simple: The expectation operator E is the same everywhere.

Now take a general statistical inference point of view on any situation that might lead to statements like (6.2) and (6.1). Then one must be prepared to take into account the fact that there are really 4 different experiments involved in these (in)equalities. The $\hat{\lambda}$'s and $\hat{\mu}$'s are random variables, but they are also connected to statistical inference in these experiments. What we know at the outset in the Einstein, Podolsky, Rosen (EPR) situation is only that some inaccessible c-variable $\pm\phi$ (possibly together with other inaccessible parameter-components) is involved in each experiment. Going from this to the observations, there are really three steps involved at each node: The components $\theta(\phi)$ are selected, there is a model reduction $\lambda = \eta(\theta)$, and finally an observation $\hat{\lambda}$. Briefly: A model is picked, and there is an estimation within that model.

6.4 Statistical Models in Connection to Bell's Inequality

Turn to general statistical theory: According to the conditionality principle (see Chapter 2), a principle on which there seems to be a fair amount of consensus among statisticians, inference in each experiment should always be conditional upon the experiment actually performed.

A motivating example for this is the following, due to Cox [48]: Let one have the choice between two measurements related to a parameter θ, one having probability density $f_1(y, \theta)$, and the other having probability density $f_2(y, \theta)$. Assume that this choice is done by throwing a coin. Then the joint distribution of the coin result z and the measurement y is given by

$$c(z)f_1(y, \theta) + (1 - c(z))f_2(y, \theta),$$

where $c(z) = 1$ if model 1 is chosen, otherwise $c(z) = 0$. Should this joint distribution be used for inference? No, says Cox and common sense: All inference should be conditional upon z.

In particular then, the conditionality principle should apply to the distribution of point estimators. Taking this into account, it may be argued that at least under some circumstances also in the microscopic case, different expectations should be used in a complicated enough situation corresponding to (6.1), and then the transition from (6.2) to (6.1) is not necessarily valid.

This is dependent upon one crucial point, as seen from the conditionality principle as formulated above: When one has the choice between two experiments, the same parameter should be used in both. How can one satisfy this requirement, say, in the choice between a measurement at a or at a'? As formulated above, the relevant parameters are λ^a and $\lambda^{a'}$ for the two experiments under choice.

Here is one way to give a solution: Focus on an apparatus which measures the spin. Make a fixed convention on how the measurement apparatus is moved from one location to the other to calibrate apparata at different locations. Then define a parameter λ which is -1 at one end of the apparatus and $+1$ at the other end. By using λ as a common parameter for both experiments under choice, the conditionality principle can be applied, and (6.1) does not follow from (6.2).

The crucial point here is that the violation of the Bell inequality is not by necessity a phenomenon that makes the quantum world completely different from the rest of the world as we know it. Regarding the term 'local realistic', ordinarily used in the derivation of Bell's inequality, a phenomenon is not in my opinion non-local if Bell's inequality is broken. But if 'realistic' means that a phenomenon always can be described by one single model, this may be a too strong requirement. Of relevance here: It has recently been pointed out [6, 7] that variants of Bell's inequality may be broken in macroscopic settings.

A more explicit argument for the correlation between spin measurements, using the prior at A connected to model reduction there, may be given as follows: At the outset the c-variable ϕ is sent to A and $-\phi$ to B. This should not be taken literally, since ϕ does not take any value, but must be interpreted in some indirect way. In any case it should be interpreted to mean that much common information is shared between the two places. The vector ϕ is in principle capable of providing an answer to any question $a \in \mathcal{A}$: Is the spin in direction a equal to $+1$ or to -1?

Since the c-variable is the same at both places, the transition probability should really be as for a single particle, taking into account the sign change at B. But this means that it can be found from Born's formula. An alternative informal argument, using explicitly the parametric model, runs as follows:

By choosing a direction a, the observer at A will have a prior on ϕ given by a probability $1/2$ on $\lambda^a = +1$ and a probability $1/2$ on $\lambda^a = -1$, where θ^a is the cosine of the angle between a and ϕ, and λ^a the corresponding reduced parameter taking values ± 1. This is equivalent to some (focused) prior on the vector ϕ which has probability $1/2$ of being $a + \epsilon$ and $1/2$ of being $-a + \epsilon$, where a is a unit vector, and ϵ is some random vector perpendicular to a which is independent of λ^a and has a uniformly distributed direction. Note that this reasonable prior on ϕ is found by just making the decision to do a measurement in the direction a at A.

Now let one decide to make a measurement in the direction b at the site B. Let $b^\perp$ be a unit vector in the plane determined by a and b, perpendicular to b. Then, taking the prior at A as just mentioned, ϕ will be concentrated on $a + \epsilon = b\cos(u) + b^\perp\sin(u) + \epsilon$ and $-a + \epsilon$ with the same a, where u is the angle between a and b, and where the sign in front of a is equal to λ^a. Hence the component of this prior for $-\phi$ along b will be $-\lambda^a\cos(u) - \epsilon \cdot b$, where the first term takes two opposite values $\pm\cos(u)$ with equal probability. The expectation of this prior component will be 0, more specifically, the component will have a symmetrical distribution around 0.

Conditionally, given λ^a, this prior component will have an unsymmetrical distribution, and there is a uniquely distributed parameter μ^b taking values ± 1 such that $E(\mu^b|\lambda^a) = -\lambda^a\cos(u)$. So, using parameter reduction to ± 1 at B, this is the distribution obtained from the model assuming a measurement in direction a at A.

There is no action at a distance here; all information is in principle contained in the c-variable $\phi = (\lambda^a, \mu^b)$.

Turning now to estimation, in general an unbiased estimator in statistical theory is a random variable whose expectation equals the parameter in question. Let now $\hat{\lambda}^a$ and $\hat{\mu}^b$ be unbiased estimators of λ^a and μ^b, respectively, so that $E(\hat{\lambda}^a|\lambda^a) = \lambda^a$ and $E(\hat{\mu}^b|\mu^b) = \mu^b$. Then

$$E(\hat{\lambda}^a\hat{\mu}^b) = E(E(\hat{\lambda}^a\hat{\mu}^b|\phi)) = E(E(\hat{\lambda}^a|\lambda^a)E(E(\hat{\mu}^b|\mu^b)|\lambda^a)))$$
$$= E(\lambda^a(-\lambda^a\cos(u))) = -\cos(u). \tag{6.3}$$

The discussion above was partly heuristic, but the above calculation using a double expectation is precise. Also, it leads to the correct answer, a well known property of electron spin, and it seems to be a way to interpret the information contained in the c-variable $\phi = (\lambda^a, \mu^b)$.

It is also important that the above discussion was in terms of a reasonable parametric model. Parameters are distinctly different from random variables, in particular from random variables located in time and space. Much of our daily life imply the use of mental models, and also some form of model simplification. Quantum theory can in some sense be said to have analogies also to this world, perhaps more than to the world of classical mechanics. This can also be related to several recent papers by Aerts and his group [7–9], where quantum mechanical modeling is used in the analysis of language and concepts.

The limitation of the way of thinking demonstrated in this section is twofold: First, the basic group needs not be the rotation group in general. Secondly, it may not be straightforward to generalise the reasoning to the case with more than two eigenvectors. Hence one can again go back to the more precise treatment in Chapter 5, where I started to build up the apparatus which is necessary to treat more general cases. Ultimately, this leads essentially to the ordinary formalism of quantum theory.

6.5 Groups Connected to Position and Momentum. Planck's Constant

I continue to insist upon keeping the distinction between ideal values of variables, that is, parameters on the one hand, and observed values on the other hand. In the statistical traditions we will continue to denote the former by greek letters. Hence let (ξ_1, ξ_2, ξ_3) be the ideal coordinates of a particle at time τ, and let (π_1, π_2, π_3) be the (ideal) momentum vector and ϵ the (ideal) energy. In this section I will not speak explicitly about observations. Nevertheless it is important to be reminded of the premise that these quantities are theoretical, and that each single of them can only be given a concrete value through some given observational scheme.

This is a general way of thinking which also seemingly may serve to clarify some of the paradoxes of quantum theory. As an example, look at the Einstein,

Podolsky, Rosen [74] situation in its original form: Two particles have position ξ^i and momentum π^i ($i = 1, 2$). Since the corresponding quantum operators commute, it is in principle possible to have a state where both $\xi^1 - \xi^2$ and $\pi^1 + \pi^2$ are accurately determined. That implies that a measurement of ξ^1, respectively π^1 at the same time gives us accurate information on ξ^2, respectively π^2. We have a free choice of which measurement to make at particle 1, but that does not mean that this choice in any way makes any influence upon particle 2. It only influences which information we are able to extract about this particle.

After this digression I continue with the single particle situation. As is well known from special relativity, the four-vectors $\xi = (\xi_1, \xi_2, \xi_3, \xi_0 = c\tau)$ and $\pi = (\pi_1, \pi_2, \pi_3, \pi_0 = c^{-1}\epsilon)$ transform according to the extended Lorentz transformation, the Poincaré transformation, which is the group which fixes $c^2 d\tau_0^2 = c^2 d\tau^2 - \sum_{i=1}^{3} d\xi_i^2$, where τ_0 is proper time, respectively $c^2 m_0^2 = c^{-2}\epsilon^2 - \sum_{i=1}^{3} \pi_i^2$, where m_0 is rest mass. This is a group of static linear orthogonal transformations of vectors together with the transformation between coordinate frames having a velocity v with respect to each other. Specifically, the coordinate vectors transform according to an inhomogeneous transformation $\xi \mapsto A\xi + b$, while the momentum vector transforms according to the corresponding homogeneous transformation $\pi \mapsto A\pi$. The group might be a natural transformation group to link to the eight-dimensional c-variable $\phi = (\xi_1, \xi_2, \xi_3, \tau, \pi_1, \pi_2, \pi_3, \epsilon)$, associated with a particle at some time τ. However, since the static rotations have representations associated with angular momenta already briefly discussed, we limit ourselves here to the group G of translations together with the pure Lorentz group.

Consider then the groups B_j given for $g_j^b \in B_j$ by $\xi_j g_j^b = \xi_j + b$, other coordinates constant, and the groups V_j given by Lorentz boosts of some size v in the direction of the coordinate axis of ξ_j for $j = 1, 2, 3$ together with the time translation group B_0 given by $\tau g_0^t = \tau + t$.

Proposition 6.5.1.   *a) These groups generate G, and they are all abelian.*

b) The groups B_j commute among themselves, the groups V_j commute among themselves.

c) Since lengths perpendicular to the direction of the Lorentz boost are conserved, B_j commute with V_k when $j \neq k$.

d) The elements of the group B_0 commutes with those of B_j ($j \geq 1$), but not with those of V_j ($j = 1, 2, 3$).

Disregarding the time translation group for a moment, it is left to consider, say, the groups B_1 and V_1 together. As is easily seen from the formula, these do not commute; explicitly, the commutator as acting upon ξ_1 is $b / \sqrt{1 - (\frac{v}{c})^2} - b$. The simplest one is B_1, which only affects the coordinate ξ_1. Hence ξ_1 is trivially permissible with respect to this group.

From the form of the Lorentz transformation

$$\xi_1 \mapsto \frac{\xi_1 + v\tau}{\sqrt{1-(\frac{v}{c})^2}}, \quad \tau \mapsto \frac{\tau + \frac{v}{c^2}\xi_1}{\sqrt{1-(\frac{v}{c})^2}} \tag{6.4}$$

and correspondingly for (π_1, ϵ), we see that ξ_1 and π_1 are not permissible when τ, respectively ϵ are variable. The linear combinations $\xi_1 - c\tau$, $\xi_1 + c\tau$, $\pi_1 - c^{-1}\epsilon$ and $\pi_1 + c^{-1}\epsilon$ are permissible. One could conjecture that these facts could be useful in a relativistic quantum mechanics, but this will not be pursued here.

Furthermore,

Proposition 6.5.2. $V^\xi = \{f : f(\phi) = q(\xi_1(\phi)) \text{ for some } q\}$ *is a subspace of* $L^2(\Phi, \nu)$ *which is invariant under the group* B_1. *The representations have the form* $U_1(g)q(\xi_1) = q(\xi_1 g) = q(\xi_1 + b)$.

But

$$q(\xi_1 + b) = \sum_{k=0}^{\infty} \frac{b^k}{k!} \frac{\partial^k}{\partial \xi_1^k} q(\xi_1) = \exp(b\frac{\partial}{\partial \xi_1})q(\xi_1) = \exp(\frac{ibP_1}{\hbar})q(\xi_1),$$

where P_1 is the familiar momentum operator

$$P_1 = \frac{\hbar}{i} \frac{\partial}{\partial \xi_1}.$$

Thus the particular group formulated above has a Lie group representation (see Appendix A.2.4) on an invariant space with a generator equal to the corresponding momentum operator of quantum mechanics. The proportionality constant $\hbar$ can be argued to be the same for all momentum components (and energy) by the conservation of the 4-vector. By similarly considering systems of particles one can argue that:

Proposition 6.5.3. $\hbar$ *is a universal constant.*

In a similar way we can show:

Proposition 6.5.4. *Time translation* $\tau \mapsto \tau + t$ *has a representation*

$$\exp(\frac{iHt}{\hbar}), \tag{6.5}$$

where H is the Hamiltonian operator.

All these operators can be connected to group representations the group G as defined above. In A. Bohr and Ulfbeck [30] it is pointed out that the Lorentz transformation (6.4) is equivalent to

$$\xi_1 \mapsto \xi_1 \cosh r_v + c\tau \sinh r_v, \quad c\tau \mapsto \xi_1 \sinh r_v + c\tau \cosh r_v, \tag{6.6}$$

where the rapidity r_v is defined by $\tanh r_v = v/c$. This makes the Lorentz boost additive in the rapidity, and all relevant operators and their commutation relations can be derived. In particular, the familiar commutation relation $X_1 P_1 - P_1 X_1 = i\hbar I$ (with X_1 being the operator corresponding to position ξ_1) holds under the approximation $r_v \approx v/c$.

The corresponding commutation relation between the time operator and the energy operator has also been derived by Tjøstheim [197] in a stochastic process setting using just classical concepts.

Starting from these commutation relations, other representations of this Heisenberg-Weil group are discussed in Perelomov [172]. A classical paper on the representation of the Lorentz group - going much further than the discussion of this Section, and also incorporating a thorough discussion of spin - is Wigner [205].

Note that the groups G_1^b and G_1^v are transitive in the case above, so there is no need for - or possibility of - a model reduction.

Proposition 6.5.5.   *In the non-relativistic approximation, ξ_1 and π_1 are permissible. The basis vectors of the Hilbert space for position ξ_1 and basis vectors of the Hilbert space for momentum π_1 are connected by a unitary transformation of the form*

$$u^\pi(\pi_1) = \frac{1}{\sqrt{2\pi\hbar}} \int \exp(\frac{i\pi_1\xi_1}{\hbar}) u^\xi(\xi_1) d\xi_1.$$

The parameters ξ_1 and π_1 can be estimated by making observations. It is natural to impose the translation/ Lorentz group upon these measurements. Thus the requirement that the basic Hilbert space also should be a representation space for the observation group, is obviously satisfied in this case.

6.6   The Schrödinger Equation

In Section 6.5 I showed that in the case of a single particle, the time translation $\tau \mapsto \tau + t$ had the group representation

$$\exp(\frac{iHt}{\hbar}), \tag{6.7}$$

where H is the Hamiltonian operator. This can be generalised to systems of several particles using an assumption of additive Hamiltonian, and assuming that the particles at some point of time were pairwise in contact, or at least so close with respect to space and velocity that relativistic time scale differences can be neglected. The operator (6.7) acts on the Hilbert space $\mathbf{H}$. In this section I work under the non-relativistic approximation.

Assume further that at time 0 a maximal measurement is done, so that the system is in some state $v_0 \in \mathbf{H}$. This means, according to my interpretation, that some experiment with reduced parameter λ^a has been done, resulting in a value λ_1^a. The construction of the Hilbert space $\mathbf{H}$ was carried out in Section 5.3. Here I will use the more elaborate, but also more context-oriented construction in Section 5.10; the starting point was then the space $\mathbf{K}^a$ of functions of sufficient observations connected to experiment a. From this we constructed $\mathbf{H}^a = A^a \mathbf{K}^a$, where A^a was given by

$$(A^a y)(\lambda^a) = \mathbf{E}^{\lambda^a}(y),$$

but where one also has $\mathbf{H}^a = C^a \mathbf{K}^a$ for some unitary operator C^a. Note that from Theorem 5.3.1 and Theorem 5.10.1, $\mathbf{H}^a = D^a \mathbf{H}$ and $\mathbf{K}^a = E^a \mathbf{H}$ for unitary operators D^a and E^a. Here $\mathbf{H}$ is the basic Hilbert space, and $D^a = C^a E^a$.

The vector v_0 corresponds to some vector w_0 in $\mathbf{K}^a$ by this unitary transformation, then to $u_0 = C^a w_0 \in \mathbf{H}^a$.

Consider now the time translation group element with step t, and assume that λ^a transforms under this group element into a new parameter $\lambda^a(t)$. By the regular representation of the time translation group, this leads to a new operator $C^{a,t}$ given by

$$C^{a,t} y(\lambda^a) = C^a y(\lambda^a(t)) = \exp(\frac{iH_1 t}{\hbar}) C^a y(\lambda^a). \tag{6.8}$$

Here $H_1 = D^a H D^{a\,-1}$ is the Hamilton operator H transformed from the basic Hilbert space $\mathbf{H}$ to the parameter space $\mathbf{H}^a = D^a \mathbf{H}$ for experiment a.

Vectors $u(t)$ in the space $C^{a,t} \mathbf{K}^a$ correspond to vectors $(C^{a,t})^{-1} u(t)$ in $\mathbf{K}^a$. In particular, then, during the time span t, we have that w_0 in $\mathbf{K}^a$ develops into

$$w_t = (C^a)^{-1} \exp(-\frac{iH_1 t}{\hbar}) C^a w_0 = \exp(-\frac{i(C^a)^{-1} H_1 C^a t}{\hbar}) w_0.$$

Transforming back from $\mathbf{K}^a$ to $\mathbf{H}$, the state vector at time t will be

$$v_t = \exp(-\frac{iHt}{\hbar}) v_0. \tag{6.9}$$

As is well known, the latter equation is just a formulation of the familiar Schrödinger equation

Theorem 6.6.1. *The time development of the state vector can be found from*

$$i\hbar \frac{\partial}{\partial t} v_t = H v_t. \tag{6.10}$$

6.7 Classical Information and Information in Quantum Mechanics

For the development of classical information theory, the pathbreaking paper by Shannon [186] has been crucial. The central problem in that paper was coding. Look upon a message source a who produces a message k with probability p_k, and assume that we wish to represent the set of messages with sequences of binary digits that are as short as possible subject to a 'decipherability' constant. Say that the length is L bits when you send a message. Let the Shannon information be defined by

$$H(a) = -\sum_k p_k \log_2 p_k. \qquad (6.11)$$

Then Shannon's noiseless coding theorem says:

Theorem 6.7.1. *a) The expected length* $\mathrm{E}(L)$ *of the message bit sequence is bounded below by* $H(a)$.

b) If we allow ourselves to code entire blocks of independent messages together, the expected number $\mathrm{E}(L)$ *of bits per message can be brought arbitrarily close to* $H(a)$.

The concept of Shannon information can also be closely connected to the concept of entropy in statistical mechanics: The entropy of a macrostate can be interpreted as the number of bits that would be required to specify the microstate of a system.

Now turn to quantum mechanics. In the ordinary formulation of quantum mechanics, and also in my approach to it, a state can be defined as being equivalent to a density matrix

$$\rho = \sum_k p_k v_k v_k^\dagger. \qquad (6.12)$$

For such a state von Neumann [163] defined his entropy by

Definition 6.7.1. $S(\rho) = -\mathrm{tr}\rho\log_2\rho$.

The logarithm can be made precise here by a suitable series expansion.

Of particular importance is the case when the density matrix is given by uncertain answers to a single question a, i.e., when p_k is the probability that the answer is k. Then the state vectors v_k in (6.12) are orthonormal. Note that since the trace of a density matrix is 1, the probabilities in (6.12) must always add to 1.

Lemma 6.7.1. *When the state vectors* v_k *are orthonormal, von Neumann's entropy is equal to the Shannon information:* $S(\rho) = H(a)$.

<u>Proof.</u>
Assume a series expansion

$$-(1-p)\log_2(1-p) = \sum_r b_r p^r.$$

Then

$$S(\rho) = \mathrm{tr}\sum_r b_r(I - \sum_k p_k v_k v_k^\dagger)^r = \mathrm{tr}\sum_r b_r(\sum_k (1-p_k)v_k v_k^\dagger)^r$$

$$= \sum_r b_r \sum_k (1-p_k)^r \mathrm{tr}((v_k v_k^\dagger)^r) = -\sum_k p_k\log_2 p_k = H(a).$$

Schumacher [181] has given a noiseless coding theorem equivalent to Theorem 6.7.1 for von Neumann's entropy which is valid both for the orthogonal and the non-orthogonal case. In Peres and Terno [173] quantum information is discussed from several points of view and also related to relativity theory.

It is interesting that there exist connections between several of the central information measures that are used in physics and in statistics. First for a continuous distribution with density $p(x)$ the Shannon information is defined in a natural way as

$$- \int p(x) \log_2 p(x) dx. \qquad (6.13)$$

This is closely related to the Kullback-Leibler entropy [139], a 'distance' measure between two densities $p(x)$ and $q(x)$:

$$G[p(x), q(x)] = - \int p(x) \log[p(x)/q(x)] dx. \qquad (6.14)$$

Finally, we have the important concept of Fisher information, defined in Section 2.12 as

$$I(\theta) = \text{Var}^\theta(s(\theta)), \qquad (6.15)$$

where $s(\theta)$ is the score function, the partial derivative of the log likelihood function.

For a simple location model with probability density $p(x - \theta)$ the Fisher information is constant and can be found as

$$I = \int \frac{p'(x)^2}{p(x)} dx. \qquad (6.16)$$

The following result was suggested by Frieden [90]:

Lemma 6.7.2. *Under suitable regularity condition the Fisher information in the location case can be found as a limit of the Kullback-Leibler information between densities that are close to each other:*

$$I = -\lim_{\delta \to 0} \frac{2}{\delta^2} G[p(x), p(x + \delta)]. \qquad (6.17)$$

Proof.
Two times use of l'Hôspital's rule under the integral sign in the definition of G. The regularity condition can be formulated in terms of an integral which is sufficiently large so that the dominated convergence theorem can be used.

I will come back to the use of Fisher information in Chapter 7. Recall from Chapter 2 the Cramér-Rao bound: No unbiased estimator of the parameter θ has variance smaller than $I(\theta)^{-1}$.

In the quantum setting one has the choice between several measurements a. In this setting Helstrom [119] defined a quantum information $I(\theta)^{-1}$ in 1967 and proved a quantum version of the Cramér-Rao bound: Based on any measurement we have that any unbiased estimator has variance smaller than $I(\theta)$. This result was further developed by Braunstein and Caves [35] and by Barndorff-Nielsen and Gill [17].

6.8   Some Themes and 'Paradoxes' in Quantum Mechanics

Here I include a very brief discussion of some familiar themes from quantum mechanics, many of which are discussed in several textbooks. Recent discussions of several points are given by Selleri [184] and Laloë [140].

Of course, much more can be said on each theme. Some of the statements below are controversial, and many are certainly too simplified. The brief statements may serve as a starting point of a discussion, however. My main concern is to point out similarities between (my version of) quantum physics and statistical modeling. Some of the statements below may be repetitions of earlier statements.

The status of the state vector.

Let us concentrate on a discrete parameter, typically multidimensional: Suppose that $\lambda^a(\phi)$ is maximal in the sense that no parameter can be connected to any experiment in such a way that λ^a is a function of this parameter. Then the operator T^a corresponding to $\lambda^a(\cdot)$ has a non-degenerate spectrum. Thus each specification $(\lambda^a(\phi) = \lambda_k^a)$ is equivalent to specifying a single vector v_k^a. I emphasise that λ^a is a parameter which is specifically connected to the experiment (or question) $a \in \mathcal{A}$.

Thus in this case the state can be specified in two equivalent ways. For a non-physicist the specification $(\lambda^a(\phi) = \lambda_k^a)$, that is, specifying all quantum numbers, is definitively simpler to understand than the Hilbert vector specification. It is easy to see that every Hilbert space vector is the eigenvector of *some* operator. Assuming that this operator can be chosen to correspond to some λ^a, it then follows that the state vector can be written as equivalent to some $(\lambda^a(\phi) = \lambda_k^a)$. A more general statement will include continuous parameters.

This also holds for a state evolving through the Schrödinger equation. But while it is true that v_t at each t is equivalent to *some* statement $(\lambda^a(t) = \lambda_k^a(t))$, the parameter $\lambda^a(t)$ will in case necessarily change with time.

Note also, of course, that in the formulae of Section 5.7 and in related results, the state vector is needed explicitly.

Collapse of the wave packet.

If we maintain that the rôle of the wavefunction is to give condensed information about what is known about one or several parameters of the system, then it is not strange that the wavefunction changes at the moment when new such information

Fig. 6.1 Schrödinger's cat.

is obtained. Such a change of state due to change of information is well known in statistics.

Superselection rules.

When parameters are absolutely conserved, for instance charge or mass, then also in conventional quantum mechanics no linear combination is allowed between the state vectors specifying linear combinations of states with different values of these parameters. This may to some extent serve to emphasise our view that a wave function makes sense only if it can be made equivalent to some statement $\lambda^a(\phi) = \lambda_k^a$.

Wigner's friend etc. This is a classical example where a person (Wigner) observes the world, and a friend also observes everything, including Wigner. In some classical interpretation of quantum mechanics it is relevant to ask which person has the correct wavefunction, Wigner or his friend.

In principle a statistical model can be formulated for a given system either excluding a certain observer (measuring apparatus) in the model, or including this observer. There is no contradiction between these two points of view in principle.

Bohr complementarity.

A limited experimental basis implies that an experimentalist must choose between measuring/ specifying the maximal parameter λ^1 or the maximal parameter λ^2. It is impossible to specify both. And knowledge of both parameters is impossible to have. As has been stated earlier, several macroscopic examples of the same phenomenon can be found.

Schrödinger's cat.

This is an imaginary situation where a cat is locked in a cage together with some radioactive substance with decay probability 1/2 during the span of time of the experiment, and some poison contained in a bottle which is broken when a radioactive particle is released. The discussion of this example concerns the state of the cat when the cage is opened. Is it half dead and half alive?

The c-variable ϕ can again be imagined to give a complete description of the whole system, including the death status of the cat. What can be observed in practice, is one of several complementary parameters λ^a, many of which include information of the death status, but some which don't. Included among the latter is the state variable developed by specifying the initial state of the radioactive source, and then letting some time go. We emphasise that our interpretation of quantum mechanics is epistemological, not ontological.

Decoherence.

When a system in a state $\lambda^a(\phi) = \lambda_k^a$ enters into an interaction with an environment with many degrees of freedom, a state involving a probability distribution over different λ_k^a-values will soon emerge.

Quantum mechanics and relativity.

Relativistic quantum mechanics is beyond the scope of the present book. However, it is well known that the use of symmetries, in particular representation theory for groups is much used in relativistic quantum mechanics and in elementary particle physics. Hence a development of the theory in that direction appears to be possible, and would certainly be of interest.

It has often been said that it is difficult to reconcile quantum mechanics with general relativity theory. While this at the moment is mere speculation, one possible explanation may be that the transformation groups in general relativity are so large that no representation theory exists. (Say, the groups are not locally compact.) In that case the formal apparatus of quantum mechanics has no place. However, in principle it might still be that the present approach base on models, symmetry, focus parameters and model reduction may prove to be useful. A much more sophisticated and more complete approach to the same question is given by Dörings's and Isham's topos formulation; see [64–67] and Chapter 9.

6.9   IIistories

The question and answer: What is λ^a?/ $\lambda^a = \lambda_k$ is represented by the state vector v_k^a or by the one-dimensional projector $v_k^a v_k^{a\dagger}$. What should be the representation be if the answer is of the less specific kind: $\lambda^a \in A$? The natural solution to this is to introduce the more general projector

$$E_A^a = \sum_{\lambda_k \in A} v_k^a v_k^{a\dagger}.$$

Such projectors can also be introduced in connection with continuous parameters, when we specify that this parameter should belong to some interval, say. Let us call a projector of this kind in general an *event*. The event 'not E' is of course specified by the projector

$$E' = I - E.$$

Griffiths [100] introduced a *history* $\mathcal{H}$ of events

$$D \to E_1 \to E_2 \to ... \to E_n \to F \tag{6.18}$$

associated with a set of times

$$t_D < t_1 < t_2 < ... < t_n < t_F.$$

The interesting thing is that under certain conditions one can introduce probabilities associated with such histories even in cases where the operators given by the various events are not commuting: When taken at different times, it may be meaningful to ask questions associated with noncommuting operators.

To make this precise, we need to introduce the time development of events. In the same way as the state vectors follow the Schrödinger equation, so do the events. For some reference time t_r let $U(t) = \exp(-iH(t - t_r)/\hbar)$ be the Schrödinger operator from equation (6.9), and define for an event E_j at the time t_j:

$$\hat{E}_j = U(t_j)E_j U(t_j)^\dagger.$$

Then the weight for the history (6.18) is defined as follows:

$$w(\mathcal{H}) = \text{tr}(\hat{E}_n \hat{E}_{n-1}...\hat{E}_2 \hat{E}_1 \hat{D} \hat{E}_1 \hat{E}_2 ... \hat{E}_{n-1} \hat{E}_n \hat{F}). \tag{6.19}$$

This turns out to be independent of the reference time.

To proceed from this, Griffiths [100] added a consistency condition, namely that the weights should add in a natural way when an event E_k is split up in a union of several disjoint events. Under this condition he proved that the weights act as probabilities: If one conditions upon the initial and final states D and F, the result acts as a conditional probability of the remaining history. Griffiths gives many examples of applications.

A stronger, but simpler consistency condition was proposed by Gell-Mann and Hartle [92]. The history approach is further discussed and related to other themes in Omnès [167].

6.10 The Many Worlds and Many Minds

In this section I will consider some rather radical viewpoints on quantum mechanics and its interpretation. They are in parts opposite to the position taken in this book: In the many worlds school one has no doubt that the state vector has an ontological interpretation. One even assumes that there exists a state vector for the whole universe, which in principle describes every possible phenomenon everywhere. Nevertheless I will give the theories some description here, both because they have many followers among physicists, and because some of the concepts involved also may be useful for understanding my approach.

In conventional quantum mechanics, as well as elsewhere in the present book, there are two different ways in which a state vector may change:

Process 1: (Cp. Theorem 5.7.4) *By measurement of a quantity with eigenstates* $v_1^b, v_2^b, \ldots$ *a state v will be changed to a state v_j^b with probability $|v^\dagger v_j^b|^2$.*

Process 2: *Continuous change of v according to the Schrödinger equation.*

From a purely ontological point of view, the process 1 is thought to be disturbing. Everett [77] set out to develop a formalism where this process was unnecessary. To this end it first became necessary to study the Hilbert space **H** for a combination consisting of the system itself together with a measuring apparatus. This Hilbert space is the tensor product of two parts, one Hilbert space for the system and one for the apparatus.

As discussed above in connection to the Bell inequality, the simplest vectors in **H** are tensor products of vectors v for the system and u for the apparatus. But the general state vector for the combined system is a linear combination of such vectors. If the combined system is in such a state w - in this book taken as a question together with the answer to that question - it is impossible to define separate states for the two parts. The following definition turns out to be useful, however:

Definition 6.10.1. *Assume that the combined system is in a state $w \in$ **H**, let $\{v_i\}$ be an orthonormal set of system states and let u be a given state vector for the apparatus. Then define*

$$v_{rel}^u = N \sum_i [(v_i \otimes u)^\dagger w] v_i \qquad (6.20)$$

as the relative state of the system for given u, where N is a normalising constant.

It can be proved that the relative state is independent of the choice of the orthonormal set $\{v_i\}$; see Everett [78]. Using the concept of relative states Everett was indeed able to build up a formalism where the discontinuous process 1 was avoided, but this was at a price: He had essentially to assume that each time a choice of measurement was done, the universe was split into a multitude of parallel universes. In fact the notion of parallel universes was not mentioned by Everett; it was introduced later by De Witt [61] and made popular by Deutsch [59]. Some physicists believe that this gives the ultimate interpretation of quantum mechanics, but many are sceptical to the idea that there at each time exists 10^{100} or more parallel universes, of which only one is observed. At least this idea is not in the spirit of Occam's razor!

I will give some discussion of a related and more recent idea, however, namely that of multiple minds; I will here follow Lockwood [148].

Fig. 6.2 Many minds.

From the mathematics of a qubit - or electron spin - it follows that a state with spin $+1$ in the z-direction can be written as

$$\frac{1}{\sqrt{2}}v^x_+ + \frac{1}{\sqrt{2}}v^x_-.$$

If we now perform a measurement of the spin in the x-direction, then according to conventional quantum mechanics a Process 1 as formulated above takes place. To avoid this, Lockwood included the state of the observer in the state vector for the system:

$$\frac{1}{\sqrt{2}}v^x_+ \otimes u_+ + \frac{1}{\sqrt{2}}v^x_- \otimes u_-. \tag{6.21}$$

For simplicity I have omitted the measuring apparatus here, pretending the observer can observe the spin component directly. Then u_+ is the state of the mind of the observer corresponding to an observation $+1$, while u_- is the state of the mind of the observer corresponding to an observation -1. With respect to u_+, the relative state of the electron is v^x_+, and with respect to u_-, the relative state of the electron is v^x_-.

According to Everett/Lockwood the world consisting of the electron and the mind of the observer follows two branches here:

$$v^x_+ \otimes u_+$$

and

$$v^x_- \otimes u_-.$$

If the observer is caught up in the entangled state (6.21) and is asked about his recollections, it will be like remembering seeing spin $+1$ *and* remembering seeing spin -1. The observer is literally in two parallel minds here!

Suppose further that we have two electrons, 1 and 2, in the singlet state discussed above in connection to Bell's inequality. Mathematically, this state can be written

$$\frac{1}{\sqrt{2}}v^{x1}_+ \otimes v^{x2}_- - \frac{1}{\sqrt{2}}v^{x1}_- \otimes v^{x2}_+.$$

In fact the direction x can be replaced by any direction here, but I will first concentrate on spin measurements in the x-direction.

The two electrons can now be far apart, and there is an observer A at electron 1 and an observer B at electron 2. According to Lockwood, we must include the states of mind of these two observers, and arrive at the complete state

$$\frac{1}{\sqrt{2}}(v_+^{x1} \otimes u_+^A) \otimes (v_-^{x2} \otimes u_-^B) - \frac{1}{\sqrt{2}}(v_-^{x1} \otimes u_-^A) \otimes (v_+^{x2} \otimes u_+^B). \qquad (6.22)$$

This is a non-local state, but it has been arrived at by local interactions: First the local interaction between the two electrons in some distant past, then the local measurement interactions between A and electron 1 and between B and electron 2.

The simplest situation is when both A and B measure spin in the x-direction. Then they can both be described as belonging to the same pair of electron/mind-worlds. As discussed earlier, the value for electron 2 is just the opposite of the value for electron 1 in this case.

A more complicated situation occurs if they decide to measure in different directions x and y. Then (6.22) must be rewritten, and it turns out that A and B find themselves in distinct pairs of electron/mind world of which they alone have deterministically got the result spin-up or spin-down. The action at a distance is just the prediction found from Born's formula. By communicating their measured results to each other, they create among themselves four such parallel worlds, tensor products of their separate worlds.

This is the starting point of the many-mind theory as it has developed from the many-world theory of Everett. In principle, one can define the possible worlds with respect to tensor products of conscious states of all human beings which figure somewhere in the universal state vectors. There are many variants of the resulting theory and several opinions about details; the reader is referred to [148] and to the discussion there.

For people outside this community, the theory seems to contain many weird elements. In particular, in the philosophy assumed in this book, the state vector is just a tool for calculating probabilities, and I see no reason to avoid the discontinuous Prosess 1. Then a basic premise for the discussion of this section drops.

Nevertheless, this discussion contains some interesting elements. First of all, the human mind is not simple. In particular, it can hold two or more ideas at the same time. This is most easily seen during a learning process, where the value taken for a parameter is not yet determined. Such learning processes will be central to the discussion of statistical inference in the next chapters.

Another concept which should be central to statistical inference, is that of focusing. Focusing on which parameter to measure was also central to the foundation of quantum mechanics as it was argued for in the previous chapters of this book. This has paved the way towards at least some elements of a common foundation, a theme which will be further developed in the last chapters of the book.

Chapter 7

DECISIONS IN STATISTICS

7.1 Focusing in Statistics

During the last 3 chapters I have tried to translate various aspects of formal Hilbert space based quantum theory into a language based upon questions and answers concerning statistical parameters, and this was taken together with symmetry aspects to give a reasonably complete basis. I did not talk too much of the experiments that could provide answers to these questions; for the most parts these experiments were assumed to be perfect.

However, in Chapters 1-3 I did discuss statistical experiments, and most of the discussion there was consistent with the concept that one finds in virtually all statistical textbooks and theoretical papers: A standard statistical experiment can be taken to be synonymous with a parametric class of probability measures. Note that this is still the central framework for the mathematical statistical culture, even though the lack of consistency of at least a naive universal application of this concept has recently been pointed out by McCullagh [156], and the whole concept has been attacked in an interesting polemic by Breiman [36].

One of several questions that I will raise in the last 3 chapters is whether the framework created from the concept of an experiment as a parametric class of probability measures is sufficient for all applications of statistics. Is a variant of Gödel's incompleteness theorem valid also here? Can one formulate questions in the statistical language which cannot be answered in the same language?

From an applied statistical point of view the answer is certainly yes. Any moderately complicated statistical investigation will raise questions that cannot be answered using formal statistics. Such questions can sometimes, but not always, be raised in terms of parameters of some chosen model. Also discussions before an experiment is performed will be conceptual, and the concepts involved can often be taken as belonging to an extended class of parameters, not unlike the concepts/parameters - or c-variables - I used in formulating the questions of quantum mechanics.

Furthermore, an issue that I will raise in these last chapters is whether applied statistics/applied macroscopic science can learn something from the above formulation of quantum theory. Since symmetry in statistics was covered in depth in Chapter 3, a main issue here will be the focusing of parameters used in asking scientific questions. Some of the examples given will be of direct relevance to quantum theory, but - even though the inspiration from Chapter 5 is there - for the most part I will discuss various aspects of focusing in general, also since I feel that this subject is not sufficiently treated in the statistical literature.

Much of this discussion can take place in relation to relatively simple models. In fact most of the models used in statistical practice are comparatively simple. On the other hand, applied scientists tend to have a more flexible attitude to the model concept than most theoretical statisticians. It is not uncommon that an applied researcher analyses his or her data using two or more related models hoping to find similar results in all the analyses. The considerations behind such choices are largely qualitative. In my opinion it is timely to start a dialogue among mathematical statisticians on the related qualitative and subject specific reasoning behind model choice and parameter choices, a reasoning which is in principle very different from doing formal calculations using some model selection criterion. Although the discussion of the previous chapters here had a completely different focus, I feel that certain aspects of such choices of model and such focusing upon model parameters can also be learned from this discussion.

In agreement with the common situation in applied research, I will stress the obvious fact that virtually all statistical models are approximations. Then model reductions may lead to poorer approximations, or in other cases the degree of approximation will be about the same. When a symmetry group on the parameter space is present in a natural way, I will assume that every model reduction is to an orbit or a set of orbits of that group, as discussed in Chapter 3. But the main issue will be focusing. A related issue is what attitude one has to the assumptions that lie behind the statistical analysis performed.

Example 7.1.1. Expressed as pairwise differences, the data $y_1, y_2, ..., y_n$ will be used in a t-test for the hypothesis $\mu = E(y_i) \equiv 0$. As is well known, the assumptions behind this test are that $\{y_i\}$ are independent and identically normally distributed, Assume now for simplicity that independence and identical distribution are reasonably safe assumptions from the context, so that the only thing we doubt, is the normal distribution. There are a large number of ways to investigate normality that has been suggested in the literature, formal tests and graphical plotting methods.

In this context, let the c-variable ϕ consist of μ together with the variance σ^2 and together with some very general measure of non-normality γ, such that $\gamma = 0$ corresponds to the normal distribution. At the very least, γ should consist of linear components β_i of theoretical moments together with distribution function values $H(x) = F(x) - \Phi((x - \mu)/\sigma)$ for all values of x, where Φ is the standard normal distribution function. Recall that a c-variable well might contain redundant information

Let the data be investigated by two scientists A and B. Scientist A choses to first perform a classical test of non-normality based upon skewness and kurtosis due to Karl Pearson, and then, if this test does not reject normality, he does the t-test. The first test he performs then depends upon β_3 and β_4. In total he asks questions to the data which depends upon an estimable set of parameters.

Scientist B has heard the the t-test is particularly sensitive to outliers, so he investigates the question of outliers by some normal plot before doing the t-test. He may choose to reject one or more outliers before the t-test. His first question to the data then depends upon $H(x)$ for large positive and negative values of x. This can be more precise by looking at the quantiles of the distribution.

In principle these two scientists can arrive at different conclusions. The reason is that they ask slightly different questions to the data, questions that in each case can be formalised using different functions of the c-variable θ.

Cases like this occur daily in applied statistics.

Example 7.1.2.

There has been a shift in attitude among applied researchers during the last decades. Earlier formal hypothesis testing with specified null hypothesis and given alternative was much used; now the great bulk of hypothesis testing is done in terms of P-values (Definition 2.4.4.). One advantage with P-values (and also a danger for people not used to this way of thinking), is that this approach involves a focusing in the testing procedure. This focusing is implied by the choice of test variable $t(y)$. Given a specific alternative or a class of alternatives, this test variable can be chosen in theory in such a way that that the power of the test is maximised, but in practice this is not always done, for various reasons.

So choose a set of feasible test variables $t(\cdot) \in T$. For $t(\cdot)$ in this set and for varying d define the c-variable as the set $\{f(d, t(\cdot), \theta) = P(t(y) > d|H_0)\}$. The outcome of the test is found by first specifying $t = t_0 \in T$ and then calculating $f(t(y_0), t_0(\cdot), \hat{\theta})$ for data y_0 and for some estimate of the extra parameter θ. Reject the null hypothesis if this number is small. Thus we again have a situation where a question is asked to the data, and the answer depends from focusing according to a function of a c-variable. Different choice of focusing may be said to correspond to complementary questions.

7.2   Linear Models

A large body of statistical applications are based upon linear models of the form

$$y = \beta_0 + \beta_1 x_1 + ... + \beta_p x_p + e, \tag{7.1}$$

where each of the variables $y, x_1, ..., x_p, e$ are n-vectors, corresponding to observations on n units, where $x_1, ..., x_p$ are taken as non-random, and where the error term e is usually assumed to be multinormal with expectation 0 and covariance matrix proportional to the identity.

A very difficult problem in many applications is to choose which x-variables to include in the model (7.1). In many cases there is a large number to choose from at the outset. In other cases one can increase the number even further by including transformations of the original variables as in polynomial regression.

From my point of view, a natural way to look upon this situation is as follows: Let first p in (7.1) be very large, so large that all x-variables of potential interest are included in the 'model'. One cannot in general use the term model in its usual sense here, since if p is larger than n, there is no way in which the parameters can be identified from the data. For similar reasons it is not reasonable in general to call the whole vector $\phi = (\beta_0, \beta_1, ..., \beta_p)$ an ordinary parameter; in the spirit of the rest of this book I will call it a conceptually defined variable or c-variable.

From this conceptually defined entity there are many ways that one can construct useful models by selecting vectors λ^a consisting of components from ϕ. The statistical literature has focused upon formal criteria like Akaike's information criterium of Mallows' C_p (; see for instance ([103])), but in concrete cases there may also be ways in which one can use subject-matter knowledge to select the model of interest.

Also there is a mathematical statistical viewpoint under which it is unreasonable to say that one model is 'right' and another model is 'wrong'. Namely, one way of looking at the whole area, is to initially let all $x_1, ..., x_p$ be random variables, in the simplest case a multinormal set of variables. Then one conditions upon a selection of these variables. Some such selections may be sensible because they lead to a reasonably small error, while others may be completely useless from this point of view, but none are 'wrong' using this way of thinking.

From a group-theoretical perspective, one can consider independent scale-transformations of each x-variable, which then induces the group of independent scale-transformations on each β_j-parameter.

Proposition 7.2.1. *This collection of scale-transformations induces a unique transformation group on the c-variable ϕ. Every set of the form $\{\phi : \beta_k = 0; k = k_1, ..., k_q\}$ will constitute an invariant set, i.e., a collection of orbits for this group.*

Proof. Straightforward: When $\beta_k = 0$, it continues to be zero under the scale group. When $\beta_k \neq 0$, it is also non-vanishing under any scale transformation.

Therefore, every model reduction of the type described above, where a set of components of ϕ are put equal to 0, constitutes an orbit or a collection of orbits of the group. Thus this is in agreement with the general model reduction strategy put forward in Chapter 3.

Another situation, to be considered in the next chapter, is when all x-variables have the same unit, so that more general linear transformations of these variables, and hence of the β-parameters may be used for model reduction.

A class of linear models with a wide variety of applications, are the analysis of variance models, where some of the x-variables can be taken as indicator variables – the main effects, and some as products of these indicators – the interactions. Again model reductions may be relevant, and it may not be an easy task in a concrete situation to find the best model reduction. Some simple rules are applicable, however, for instance: If a main effect is deleted, then all interactions containing that main effect should be deleted.

There are many textbooks on linear models; in my setting it is natural to refer to Searle [183]. In this book overparametrised 'models' are used as a rule; in order that it shall be possible to solve several real, maximal linear models under the same umbrella, solutions are sought using generalised inverses.

7.3 Focusing in Decision Theory

Decisions in statistical inference were briefly discussed in Section 1.4. In ordinary inference theory one always imagines a fixed model behind any decision, and then the typical decision from data will result in an estimated parameter or the rejection of a hypothesis, say.

Combining this with a model reduction of the type discussed in the previous section, leads to a two-step decision procedure of the type discussed in Section 1.4. There, such a decision procedure was required to be unbiased in the sense that the decision taken at some given step should not influence decisions taken at later steps. In the present case, this is true if and only if the reduced model may be considered to be the true one. Then the decisions taken about the parameters in this model are decisions about the true parameters, and these decisions are not influenced by the way the reduced model was obtained.

Like all statements about hypothetical models, the statements above must be qualified somewhat in practice. In the real world, all models are approximations. Thus it is a simplification to talk about true models and true parameters. The parameters of the approximate model are words that can be used to express scientific statements of the type: The expected yield was estimated to be 15 kg per square meter; the hypothesis that the probability of recovering under a given treatment is higher than 0.99, was rejected.

A totally different set of situations for which focusing in decision situations is relevant, is when one is interested in testing several statistical hypotheses within the same model. There is a large literature on multiple testing of relevance to situations like this; see for instance Miller ([153]). I will not discuss the choice between the different options here, but only point to the fact that there exist welldefined cases where this literature is not relevant, and where one should rely upon single tests. These are just the cases where one can define a meaningful focus corresponding to each single test. Let us first briefly discuss a situation where multiple testing does have relevance.

Example 7.3.1. In a completely randomized experiment – meaning that one from nk plots selects randomly n plots for each variety – the yield of k varieties of wheat are compared. In most situations one will use a one-way analysis of variance model

$$y_{ij} = \mu_i + e_{ij}, \tag{7.2}$$

where μ_i is the expected yield of variety i, y_{ij} is the observed yield of this variety on plot j and e_{ij} are the error terms, usually assumed independent $N(0, \sigma^2)$.

The first step in the statistical analysis is to perform an F-test of the overall hypothesis $\mu_1 = \mu_2 = ... = \mu_k$. Usually this hypothesis is rejected, and the question is then where the differences lie. This leads to $k(k-1)/2$ partial hypotheses of the form

$$\mu_i = \mu_{i'}.$$

Testing these as single hypotheses will usually lead to a large number of false rejections just because of the large number of tests performed. This is usually compensated for by using some multiple test procedure, the simplest device being the Bonferroni test: Use the level $2\alpha/k(k-1)$ instead of α in each single test.

Example 7.3.2. Consider an experiment with k treatments modeled by the one-way analysis of variance model (7.2), but where the treatments are more structured than the varieties of the previous example. We assume in particular that it is meaningful to define an orthogonal set of k contrasts $\lambda_i = \sum_j c_{ij}\mu_j$. Such a linear combination λ_i is a contrast if $\sum_j c_{ij} = 0$ (; see Section 2.10). The set above set of contrasts is orthogonal if $\sum_j c_{ij}c_{i'j} = 0$ for $i' \neq i$. An example of the first couple of contrasts in an orthogonal set might be $\lambda_1 = \mu_1 - \mu_2$, $\lambda_2 = (\mu_1 + \mu_2)/2 - (\mu_3 + \mu_4)/2$, $\lambda_3 = \mu_3 - \mu_4$. In such cases the questions: 'Is λ_i equal to 0?' imply essentially different focus for different i.

The point is then: In all applications, the hypotheses $\lambda_i = 0$ are tested separately for different i. There is never any question of involving the multiple testing philosophy like in Example 7.3.1. The reason is that orthogonal contrasts always imply essentially different problems, and testing one such contrast involves a particular choice of focus.

Example 7.3.3. Consider a linear model

$$y = \beta_0 + \beta_1 x_1 + ... + \beta_q x_q + e,$$

where it is assumed that all parameters can be estimated. Then in all applications the hypotheses $H_j : \beta_j = 0$ are tested separately, and there is no question of including any multiple testing consideration. The reason is again that these hypotheses imply essentially different problems, and testing one such hypothesis involves a particular choice of focus.

Decisions in Statistics 189

Let the above two subject areas: model selection and focusing on a parametric hypothesis to test, stand as examples of focusing when making decisions related to statistical modelling. In both cases one starts with a parameter or other c-variable ϕ, and ends up with making a decision which in an essential way depends on a parameter λ, a function of ϕ. Thus we end up with a general situation which is related to the discussion of quantum mechanics in Chapter 5. Of course this does not mean that the two situations are identical, only that they both belong to a more general situation related to focusing of parameters, and thus bear some resemblance. A specific set of assumptions in Chapter 5 was the dependence on certain group theoretic formulations. A specific aspect of the situation of the present section is that decisions also depend upon data. However, these facts do not set the two areas completely apart. It was mentioned above that group theoretical considerations could be useful in model reduction, and for instance in Section 5.7 it was illustrated that the use of data is absolutely relevant to quantum mechanical predictions.

Finally, let us return to the one-way analysis of variance situation with contrasts.

Example 7.3.4. Assume that there are two set of orthogonal contrasts $\lambda^a = \{\lambda_i^a\}$ and $\lambda^b = \{\lambda_i^b\}$ given by $\lambda_i^a = \sum_j c_{ij}\mu_j$ and $\lambda_i^b = \sum_j d_{ij}\mu_j$. Inference on λ^a and λ^b will constitute complementary questions. To fit into the general framework, λ^a and λ^b are each maximally estimable functions of $\phi = (\mu_1, ..., \mu_p)$, albeit one-to-one functions in this case.

7.4 Briefly on Schools in Statistical Inference

The conditionality principle and the likelihood principle were formulated in Section 2.13. The conditionality principle roughly states that the experimental evidence concerning a statistical parameter is the same when one conditions upon an observation whose distribution does not depend upon the parameter. The principle was mentioned two times during my discussion of quantum mechanics: during the derivation of Born's formula in Section 5.5 and in the discussion of Bell's inequality in Section 6.4.

The likelihood principle states that all information about a parameter is contained in the likelihood function for this parameter, given the observations. It is shown in ([25]) that the likelihood principle can be formally derived from the conditionality principle together with a sufficiency principle formulated there. It is also argued in ([25]) that the likelihood principle in a natural way leads to a Bayesian approach to statistics. The Bayesian paradigm for statistical inference in general was discussed in Section 2.6.

This paradigm for statistical inference is conceptually rather clean and simple, and it has gained increasing popularity during the last years. There is no doubt that it can be useful in certain situations, in particular when the goal is to quantify our belief about the statistical parameter of the model. But in my view the Bayesian paradigm can not be taken as a principle embracing all cases where statistics is used in practice. This is an important point, both for the practical use of statistics as a tool, and for the general view, underlying much else in this book, that it is difficult to derive all tools that should be of relevance to empirical science from a small number of formal mathematical axioms. A number of arguments for this view in the present case, are as follows:

1. As argued for by examples in Section 2.13, the conditionality principle should not be taken to have universal validity.

2. This is connected to the fact that most models in practice are approximations.

3. It is also connected to the fact, underlying my view both on quantum theory and on statistical inference, that scientific questions in most applications involve focused questions. Such a focused question may involve a function of the model parameter, but it may also indirectly involve other unknown quantities, such that all these together constitute a c-variable.

4. The concept of information about a parameter is left undefined both in the conditionality principle and in the likelihood principle. Even though this is done deliberately to make the concept general, it is not immediately suitable for situations where we ask focused question about a partial parameter, at least when this also involves some external unknown information. Note that the formulation of the model itself often will constitute such unknown information.

5. The Bayesian way of thinking makes extreme use of the probability concept as the only mathematical tool. Even all prior knowledge is formulated as a probability distribution. In some cases this will be unnatural. In general probability theory is important in statistics, but it is often relevant in applications to consider other tools, like focusing of interest, like linear algebra in linear models or like symmetry as expressed through group theory.

6. Both testing of hypotheses and confidence intervals (Section 2.4) will often be suitable tools for answering focused questions. These tools are not covered under the Bayesian umbrella.

7. As mentioned in the next section, the analysis of designed experiments make use of a randomisation theory, involving ideas that are not Bayesian.

8. As argued by leCam [25], the argument from the likelihood principle to a universal Bayesian viewpoint may be a weak one.

9. The Bayesian paradigm is a rather formal mathematical theory derivable from a few axioms. By the extension of Gödel's theorem discussed in Section 4.2 one might expect that the whole way of thinking -when applied to the empirical world - is incomplete in some way.

The conclusion from all this is that I would recommend a Bayesian tool kit in some cases when it is natural, in particular when a natural prior distribution can be defined, and a classical statistical tool kit in other cases. In some cases it might be appropriate to carry out both analyses on the same data set, as complementary approaches, and hope that the two will lead to qualitatively similar results. If this is not the case, a closer examination of the problem, the model and the data might be needed. In general a statistical analysis is not like solving a mathematical problem, where one every time finds a unique solution. It can indeed be fruitful to look upon data analysis using different languages, and then try to find a synthesis.

Note in particular that there does exist general cases where the two forms of statistical analyses lead to exactly the same result, even though the interpretation is completely different; see Section 3.6. This last case, where a transitive group is defined on the parameter space, is also a case where a natural prior can be found as the right invariant measure, and it is furthermore the case where the best equivariant estimator under quadratic loss is equal to the Bayes estimator; see Corollary 3.5.5.

7.5   Experimental Design

In general, statistical analysis is appropriate in two cases, when one has observational data, and when the data stems from a designed experiments. The theory of experimental design has played an important historical rôle in the development of statistics, and currently it is essential to a lot of applications within medicine, biology, agriculture and industry. Nevertheless, courses in experimental design have sometimes been lacking from university curricula, or, if such a course has been included, it has played a rather remote rôle compared to the main courses devoted to the development of models based on probability theory and the statistical analysis developed from these models.

A few aspects of experimental design theory have been sketched in Section 2.14, and a fuller discussion of many topics of interest can be found in Box, Hunter and Hunter [34]. From my point of view, the following is important: Empirical investigations in general can be fruitfully carried out by first posing focused questions and the seeking answers through data. Experimental design can be looked upon as the science of posing questions in an intelligent way.

Specifically, the discussion in Section 2.14 on a c-variable-based general approach is important to the philosophy of this book. This is not standard material in experimental design books, but it can be easily developed further into posing focused questions in the form of factorial experiments, either full or fractional [34].

Example 7.5.1. Consider a 2^{5-1}-experiment, that is, an experiment with 5 factors, each of 2 levels, but where one cannot afford a full replicate: Only $2^4 = 16$ runs are made, but this is done in a balanced way such that all main effects ν_j can be estimated, and in addition two factor-interactions, being aliased with three factor-interactions. Here, each ν_j is the difference between means of the treatment parameter θ_t, difference between high and low level for the factor j. This implies that ν_j is an estimable function of the c-variable ϕ which was constructed there.

Since one does not have a full replicate, the further statistical analysis of the 2^{5-1} experiment is non-standard. One way to proceed is to do a normal plot in order to pick active (important) factors, and then perhaps neglect the others in the subsequent analysis. There are several choices which must be done in carrying this out, and different scientists A and B may perhaps be lead to different conclusions, as in Example 7.1.1. By careful construction the procedure of A, respectively B may be formulated as questions using parameters which are functions of the c-variable in Section 2.14.

Now turn to the common case of several replications. The statistical analysis from designed experiments are often done using the randomisation distribution, that is, no statistical model is used, but tests are performed using as reference the distribution obtained from the randomisation of the experiment itself. This is an area of statistics which is very far from the Bayesian paradigm.

More advanced topics in experimental design, say, discussions of incomplete blocks, uses group representation theory, and is thus mathematically not far from the group approach that I used in Chapter 5 when deriving quantum mechanics. The strata for randomised experiments discussed by Bailey [15] is a concrete example of a case where such a connection may be seen. Of course there is no simple direct line from this to quantum mechanics, but in both areas one finds an extensive use of eigenstuctures of operators. In several parts of advanced experimental design one will find something similar, for instance in the discussion of generally balanced designs, where a main topic is the simultaneous diagonalisation of operators; see Nelder [161] and references there.

7.6   Quantum Mechanics and Testing of Hypotheses

In this section we continue the discussion of Section 5.7, and recall some of the results there for convenience.

In the approach of this book, the quantum states v_k^b are equivalent to a question: What is the value of the parameter λ^b? together with the crisp answer: $\lambda^b = \lambda_k^b$. In the spin 1/2 case this will correspond to a chosen direction b, and the question is about what the spin component in this direction is. The answer can be $+1$, corresponding in ordinary quantum mechanical terms to a certain eigenvector v_1^b, or -1, corresponding to the orthogonal eigenvector v_2^b.

Many textbooks in quantum mechanics discuss the Bloch sphere representation of this result:

$$v_1^b v_1^{b\dagger} = \mathcal{E}_1(b) = \frac{1}{2}(I + b \cdot \sigma) \tag{7.3}$$

$$v_2^b v_2^{b\dagger} = \mathcal{E}_2(b) = \frac{1}{2}(I - b \cdot \sigma), \tag{7.4}$$

also mentioned in Chapter 5 (see equations (5.22) and (5.23).)

In this section I will look at a simple hypothesis testing problem with fixed level α and fixed power β for a parameter λ^b taking two values ± 1, like in Section 5.7. For simplicity I choose the test statistics t such that it also takes the two values $+1$ and -1, which always is possible. The level and power requirements then take the form

$$P(t = -1|\lambda^b = +1) = \alpha,$$

$$P(t = +1|\lambda^b = -1) = 1 - \beta.$$

This means that we are testing the hypothesis $H_0 : \lambda^b = +1$ against the alternative $\lambda^b = -1$, and that the hypothesis is rejected when $t = -1$.

Note that the result of such a test still must be considered as a state of the question/answer type, albeit in a more advanced form: The question is given by the three-vector b and the two predetermined error probabilities α and $1 - \beta$. The answer is given by $\lambda^b = +1$, say, which is the conclusion we arrive at if we observe $t = +1$. So any specification of the state must involve all the quantities b, α, β and the answer ± 1.

Such a specification can be done in a natural way if we introduce the concept of effect introduced in Chapter 5 in connection to the discussion of the Born formula. Specifically, say that we have done the experiment and reported the value $t = +1$. Then we may specify the resulting state by weighting the respective effects (7.3) and (7.4) according to the relevant probabilities for the given result. This gives, as in equation (5.28):

$$\mathcal{E} = (1 - \alpha)\frac{1}{2}(I + b \cdot \sigma) + (1 - \beta)\frac{1}{2}(I - b \cdot \sigma) = \frac{1}{2}((2 - \alpha - \beta)I + (\beta - \alpha)b \cdot \sigma). \quad (7.5)$$

Lemma 7.6.1. *From the operator $\mathcal{E}$ above one can determine uniquely b, α and β. Conversely, these three entities determine $\mathcal{E}$ as above for the case $t = +1$.*

This concept of effect for testing of hypothesis is consistent with the general definition of effect given in definition 5.7.2. In the present two-dimensional (spin) situation such an effect can always be written $\mathcal{E} = \frac{1}{2}(rI + cb \cdot \sigma)$; thus we have $r = 2 - \alpha - \beta$ and $c = \beta - \alpha$.

In the (r, c)–plane, the effects are limited to the triangle with corners $(0, 0)$, $(2, 0)$ and $(1, 1)$. But note that b can be replaced by $-b$ depending on the outcome, so the triangle obtained by taking the mirror image around the line $c = 0$ is also relevant. The first triangle corresponds exactly to the limitation imposed by the hypothesis testing interpretation discussed above:

The limitations imposed by the triangle are $c \geq 0$, corresponding to $\beta \geq \alpha$, and $c \leq r \leq 2 - c$, corresponding to $\alpha \geq 0$ and $\beta \leq 1$. The bottom line $c = 0$ corresponds to $\beta = \alpha$, a case where there is no information in the reported result. The right boundary corresponds to $\alpha = 0$, and the left boundary to $\beta = 1$. And at the top of the triangle, where $\alpha = 0$ and $\beta = 1$, both error probabilities are 0, and we get the pure states on the Bloch sphere.

A simple hypothesis problem can be inverted by exchanging hypothesis and alternative.

Lemma 7.6.2. *Assume also that the reported result is opposite of what we had above, that is, -1 instead of $+1$. Then the operator will be:*

$$\mathcal{E}_2(r, c, b) = \frac{1}{2}((2 - r)I - cb \cdot \sigma) = I - \mathcal{E}.$$

So what do we get from all this? Obviously there is a correspondence between concepts of quantum mechanics and concepts corresponding to a simple hypothesis testing situation. The result of importance, however, is that this correspondence implies that one sometimes can use the probability results from quantum mechanics in a macroscopic situation.

The point here is that one can learn something a priori from one performed experiment about the potential result of another experiment. I will discuss in detail one particular example here. The importance of this example lies in the fact that it possesses so much symmetry that the results of quantum mechanics, in particular Born's formula, is of relevance.

Example 7.6.1.

Four drugs A, B, C and D are being compared with respect to the expected recovery time μ they induce on patients with a certain disease. Since these expectations are difficult to estimate, one concentrates on getting information on the difference between each μ and the others, for instance

$$\lambda^A = \text{sign}(\mu_A - \frac{1}{3}(\mu_B + \mu_C + \mu_D)).$$

I will not go into detail with the experimental design here, but assume that there is an efficient design, say, of an incomplete block type, where accurate information can be obtained about one or a few such λ's.

Assume that we from some experiment have obtained very accurate information that $\lambda^A = +1$, and that this is the only information we have. Then we want to perform a new experiment in order to test a hypothesis that λ^B also is $+1$. Can we get any prior information about the result of this from the first experimental result? Informally, since μ_A is subtracted in the expression for λ^B, we should expect a probability less than $1/2$ that $\lambda^B = +1$.

Our main assumption now is that there is a complete symmetry between the 4 binary parameters $\lambda^A, \lambda^B, \lambda^C$ and λ^D. Permutational symmetry can always be imagined as imbedded in some rotational symmetry. Here we can consider rotation in 3-space, looking upon a regular tetrahedron in this space. The perpendiculars from the corners A, B, C and D of that tetrahedron to the opposite side can then

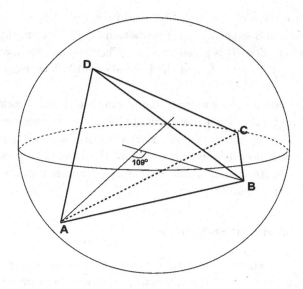

Fig. 7.1   Tetrahedron with two perpendiculars.

be taken to represent the parameters $\lambda^A, \lambda^B, \lambda^C$ and λ^D, respectively. From what is known about regular tetrahedrons, the angle between two perpendiculars is approximately $109°$, with a cosine equal to $-1/3$.

Proposition 7.6.1. *Assume 1) that the transition probability from $\lambda^i = +1$ to the conclusion $+1$ for a simple test of $\lambda^{i'}$ with level α and power β is the same for each choice of different i and i' among A, B, C and D; 2) that the same result holds for arbitrary sets of 4 drugs A', B', C' and D'. Then this transition probability is*

$$1 - \frac{1}{3}\alpha - \frac{2}{3}\beta. \tag{7.6}$$

For the special case of an ideal experiment with $\alpha = 0, \beta = 1$, the probability is just 1/3.

Proof. From assumption 1) we have the symmetry needed for the Born Theorem (Theorem 5.5.1) under the permutation group. By assumption 2) we have the same symmetry for every tetrahedron as sketched above, and this gives symmetry under the whole rotation group. This leads to the same situation that we have for the spin 1/2 case, and for the ideal case the transition probability is given in Proposition 5.6.2 with a and b representing the two directions as given by the perpendiculars in the tetrahedrons. This gives for an ideal experiment:

$$P(\lambda^b = +1 | \lambda^a = +1) = \frac{1}{2}(1 + \cos(a, b)) = \frac{1}{2}(1 - \frac{1}{3}) = \frac{1}{3},$$

and then by symmetry $P(\lambda^b = +1 | \lambda^a = -1) = 2/3$.

The general case then follows from (7.5) and from the additivity of generalised probability discussed in Section 5.5 (see Proposition 5.5.2). Specifically, we want to find $\pi(\mathcal{E}(\alpha, \beta))$, where $\mathcal{E}(\alpha, \beta)$ is given by (7.5), Then the results above show that $\pi(\mathcal{E}(0,1)) = \frac{1}{3}$ and $\pi(\mathcal{E}(0,0)) = \frac{2}{3}$, and the general result follows from linearity.

An interesting point is that a natural noninformative Bayesian prior will give a different answer here for the ideal experiment. In fact, the same answer will be given by any identical and independent normal prior on μ_A, μ_B, μ_C and μ_D. A numerical integration gives the answer 0.43 instead of 1/3 for this case. The question is if this normal assumption is stronger than the symmetry assumption made above. This will depend on the circumstances.

7.7   Complementarity in Statistics

Recall Definition 1.5.1, where two experiments are called incompatible if they cannot be performed at the same time. Many physicists will use the term complementary here, and in this section I will follow this terminology, even though Accardi has proposed a stronger definition of complementarity, as mentioned in Chapter 1. I will call this latter property strong complementarity: Two experiments A and B are strongly complementary if any experiment resulting in a fixed value for the variable of A must result in a uniform distribution of the variable of B.

I will now try to look at the relevance of these concepts in a statistical context. The first item we must discuss is what we should mean by experiment and by variable in this connection. Recall from Chapter 1 that I have included the choice of focus in the concept of experiment: Every experiment begins with a question.

In particular, the choice of a statistical model in this case is empirically determined, and the estimation of a parameter within that model must be considered as being related to two different experiments. In regression analysis one routinely uses complementary tools for these tasks: Least squares calculations for estimating the parameters of a given model and residual plots for investigating if the model can be considered to be correct, given the data. If the model is rejected by the data, the estimates under the model are irrelevant; in this sense the two experiments must be said to be incompatible.

When testing parameters, the focus is more specific, as shown in Examples 7.3.2 and 7.3.3. For a single test one can consider a c-variable which is 1 when the hypothesis should be rejected, otherwise 0, or eventually some distance from the hypothesis, perhaps in one specific direction. Two such tests must usually be considered incompatible in the sense that the result of one test anwers a completely different question than the other test. In most cases, also, the two tests will not be independent, so that the result of a simultaneous test can not be predicted from the results of the two single tests.

A special role in the connection between statistics and quantum mechanics is played by

The Cramér-Rao inequality

$$Var^\theta(\hat{\theta})Var^\theta(s(x,\theta)) \geq 1,$$

where $s(x,\theta) = \frac{\partial}{\partial\theta}\ln f(x,\theta)$, was formulated in Section 2.12.

In the present situation $\hat{\theta}$ and $s(x,\theta)$ play the role as two complementary variables, and these may be connected to two incompatible, essentially different questions. Of course $\hat{\theta}$ is connected to the estimation of θ at some point. The score $s(x,\theta)$ may be connected to a one-sided test of the hypothesis $\theta = \theta_0$. Specifically, the score test, based on the score statistic

$$z = \frac{s(x,\theta_0)}{\sqrt{I(\theta_0)}}$$

is a good approximation to the locally most powerful test in this case, and it can also be related to the asymptotic theory of such tests [50].

Chapter 8

MULTIVARIATE DATA ANALYSIS AND STATISTICS

8.1 Introduction

In recent years there has been a large scientific activity connected to the development of data-analytic methods which at first sight seem to have little or no connection to traditional statistics. Some of these methods have been originated within the machine learning community; some are separate developments. Many of the methods are discussed - to some extent from an ordinary statistical perspective - in Hastie, Tibshirani and Friedman's monograph [103].

In this book I have to a large extent argued for non-traditional ways of looking at ordinary statistics. This has been related to my views upon quantum mechanics, and the viewpoint that there can be seen a connection between the two areas related to focusing upon functions of c-variables, to model reduction, to symmetry and to the concept of complementarity. A remaining question is to which extent this connection can be extended to the other data-analytic methods.

I will give a brief review of some of the other methods in Section 8.7 below, but in his book I will concentrate on one of them, namely the partial least squares (PLS) method originated by Herman Wold, partly under the name 'soft modelling'. Today the partial least squares culture may be considered as split into two directions, a class of prediction methods employed in chemometrics and related fields on the one hand, and soft models with many variables used in economical and social science investigations on the other hand. There has been relatively little communication between these areas and ordinary statistics. In Section 8.2 I will introduce the partial least squares algorithm used in prediction problems and show later that it can be connected to a 'hard' statistical model, which can most naturally be seen as a reduced model with the kind of model reduction discussed in Chapter 3 above.

Note that one element of the statistical paradigm: The distinction between parameters and observations, will be crucial to the conclusion of this discussion. This distinction has been looked upon with scepticism by the PLS-community, who tend on the contrary to emphasise the 'softness' in their own way of thinking. A similar attitude can be found in some other data-analytic communities, and it has spread to certain users of data analysis. Our view is clear here: By emphasising model reduction as one of the main concepts upon which the connection between paradigms is built, the importance of the distinction between parameters and observations is enhanced rather than reduced. As in the earlier chapters of this book, however, I will at times allow myself to look upon parameters as more than just indices of probability distributions for data.

The motivation for introducing this chapter here, is the following: In the previous chapters, a new foundation of quantum mechanics was introduced as a statistical theory under symmetry and focusing. Thus a link between different cultures was emphasised, with an extended statistical culture as a basis. In this chapter we will look at a new culture, the PLS-culture, and we will show that this can be derived from an extended statistical theory, where the extension is based upon symmetry and model reduction. The conclusion from all this is that there exists an extended statistical culture which embaced both quantum mechanics and the data analysis in PLS. Thus we have at least two indications of a possible unity of scientific methodology.

8.2   The Partial Least Squares Data Algorithms

The least squares method from the ordinary regression model $y = X\beta + \epsilon$ boils down to a projection method: The prediction of y is given by $\hat{y} = X(X'X)^{-1}X'y$ and the residual vector is $y - X(X'X)^{-1}X'y$. A stepwise version of this can be seen as the motivation behind the partial least squares (PLS) algorithm.

A crucial element of the algorithm is that it is assumed to be stopped after a certain number of steps; this number is most often determined by cross validation.

There are variants of the PLS algorithm, but these are equivalent; see [104] and Appendix A4. I will formulate below the original algorithm, also the most common one in use today.

Algorithm 8.2.1. *Start with data (X, y) of dimension $n \times (p+1)$ as in ordinary regression. To simplify, we assume that the X-matrix and the y-vector are centered, that is, the mean has been subtracted in each column. Put $E_0 = X$ and $f_0 = y$. Then compute in steps for $k = 1, ..., a$:*

$$t_k = E_{k-1}w_k \quad \text{with weights } w_k = E'_{k-1}f_{k-1}, \tag{8.1}$$

$$p_k = E'_{k-1}t_k/t'_k t_k = X't_k/t'_k t_k, \tag{8.2}$$

$$q_k = f'_{k-1}t_k/t'_k t_k = y't_k/t'_k t_k, \tag{8.3}$$

$$E_k = X - t_1 p_1' - \dots - t_k p_k', \tag{8.4}$$

$$f_k = y - t_1 q_1 - \dots - t_k q_k. \tag{8.5}$$

The equalities in (8.2) and (8.3) may be justified by first noting that the vectors t_k are orthogonal (cp. Lemma 8.2.1), then using (8.4) and (8.5).

The interpretation of the first crucial part of this algorithm is as follows: In equation (8.1), the latent variables or scores t_k are constructed as linear combinations of the x-variables from the previous steps. What can be discussed in connection to this equation, is if the choice of weights w_k by necessity is the best possible. It has been pointed out [129] - see Appendix 4 - that PLS with this choice can be motivated by the property that the sample covariance between y and t_1 is maximised in the first step, and that a similar property holds in later steps. This may certainly be seen as a nice property, but it is a property that is connected to each single step. Any satisfactory optimality property of a prediction method should be connected to the method as such, not to the substeps in the construction of the method.

But anyway, accepting this step, in equations (8.2) and (8.3), loadings p_k and q_k are constructed by least squares fit. Finally, in equations (8.4) and (8.5), new x- and y-variables are computed as residuals in latent structure equations.

If now $x_0 = (x_{01}, \dots, x_{0p})'$ is a set of x-measurements on a new unit, one defines $e_0 = x_0 - \bar{x}$ with $\bar{x} = (\bar{x}_1, \dots, \bar{x}_p)'$ and then new scores and residuals consecutively by

$$t_{k0} = e_{k-1}' w_k, \quad e_k = e_{k-1} - t_{k0} p_k, \tag{8.6}$$

corresponding to equations (8.1) and (8.4). Then finally the y_0-value corresponding to x_0 is predicted in step a by

$$\hat{y}_0 = \bar{y} + \sum_{k=1}^{a} t_{k0} q_k = \bar{y} + \sum_{k=1}^{a} t_{k0} (t_k' t_k)^{-1} t_k' y. \tag{8.7}$$

The number of components a is usually determined by cross-validation, but other options have also been proposed [116, 124, 150]. Another possibility, if there is enough data, is to use an independent test set.

Note that (8.4) and (8.5) give a decomposition of (X, y) into latent variables and residuals. Plots of scores t_j and loadings p_j are much used in the applications of this method. Several mathematical properties of the variables of the algorithm can now be found.

Lemma 8.2.1. *The scores t_k are orthogonal.*

<u>Proof.</u>
This follows from equations (8.1), (8.2) and (8.4).

Proposition 8.2.1. *(a) The weights satisfy the recursion relation*

$$w_{k+1} = s - SW_k(W_k'SW_k)^{-1}W_k's, \tag{8.8}$$

with $W_k = (w_1, ..., w_k)$, $S = X'X$ *and* $s = X'y$.
(b) The weights w_k *are orthogonal.*

Proof. (a) here can be seen by using an equivalent algorithm - see Appendix 4. Given (a), the orthogonality (b) follows by premultiplying (8.8) by W_k'. (Another proof of this is given in [129], where also other orthogonality relations are discussed).

A further result is most easily proved by the above mentioned alternative, equivalent algorithm (Appendix 4) is that:

Proposition 8.2.2. *The PLS regression vector with a components can be written*

$$b_a = W_a(W_a'SW_a)^{-1}W_a's. \tag{8.9}$$

A final important result is then:

Proposition 8.2.3. *The weights* $w_1, ..., w_k$ *span the same space as the Krylov sequence:* $s, Ss, ..., S^{k-1}s$.

Proof.
We have $w_1 = s$. Use induction in k together with (8.8) to show that each w_k is a linear combination of $s, Ss, ..., S^{k-1}s$. It is left to prove that this latter set of vectors is linearly independent. It is shown in the next section that this holds as long as k is smaller than the number of different eigenvalues of S. If this latter condition does not hold, the algorithm will stop in the sense that the next w_k vanishes.

Note that from this we have that W_a in equation (8.9) may be replaced by $V_a = (s, Ss, ..., S^{a-1}s)$, giving a simple explicit form for the PLS regression vector.

A few exact results have been proved about PLS regression. In de Jong [58] it is shown that PLS shrinks in the sense that one always has:

$$|b_1| \leq |b_2| \leq ... \leq |b_p|.$$

A similar shrinkage result was also proved by Goutis [98]. Also, de Jong [57] proved that with the same number of components, PLS will always give a higher coefficient of determination R^2 than principal component regression PCR.

Thus the PLS algorithm implies several nice mathematical properties for the variables involved, but its motivation remains heuristic. It has also been argued in [110] that for theoretical reasons it is very unlikely that the PLS method gives optimal prediction in any reasonable sense. This has recently been confirmed by Cheng and Wu [42], who constructed concrete improvements of the method and verified numerically that these were indeed improvements. Nevertheless, one can argue that PLS in essence is a good idea, something that can be seen by going from the data algorithm to the corresponding parameter algorithm.

8.3   The Partial Least Squares Population Model

The PLS algorithm was shown above to connect to properties of the empirical covariance structure (S, s). To study the theoretical properties of this algorithm, it is natural to replace S here by the theoretical x-covariance matrix Σ, and replace s by the (x, y)-covariance vector $\zeta = \text{Cov}(x, y)$. In effect this is done by replacing Algorithm 8.2.1 by

Algorithm 8.3.1. *Let x be a random p-vector with expectation μ_x, and let y be a random scalar with expectation μ_y, and let the covariance structure be as above. Define starting values for the x-residuals ϵ_k and y-residuals f_k:*

$$\epsilon_0 = x - \mu_x, \tag{8.10}$$

$$f_0 = y - \mu_y. \tag{8.11}$$

Then compute in steps for $k = 1, ..., a$:

$$\tau_k = \epsilon'_{k-1}\omega_k \quad \text{with weights } \omega_k = \text{Cov}(\epsilon_{k-1}, f_{k-1}), \tag{8.12}$$

$$\pi_k = \text{Cov}(\epsilon_{k-1}, \tau_k)/\text{Var}(\tau_k), \tag{8.13}$$

$$\phi_k = \text{Cov}(f_{k-1}, \tau_k)/\text{Var}(\tau_k), \tag{8.14}$$

$$\epsilon_k = \epsilon_{k-1} - \pi_k \tau_k, \tag{8.15}$$

$$f_k = f_{k-1} - \phi_k \tau_k. \tag{8.16}$$

It is clear from (8.10), (8.11), (8.15) and (8.16) that this algorithm at each step a gives a bilinear representation:

$$x = \mu_x + \pi_1 \tau_1 + ... + \pi_a \tau_a + \epsilon_a, \quad y = \mu_y + \phi_1 \tau_1 + ... + \phi_a \tau_a + f_a. \tag{8.17}$$

From the usual regression argument using (8.13)-(8.16), the residual ϵ_k will be uncorrelated with τ_k; hence all the scores $\tau_1, ..., \tau_a$ will be uncorrelated. As a result, we have the alternative expression $\omega_k = \text{Cov}(\epsilon_{k-1}, y)$, which can be used in (8.12), and we can replace ϵ_{k-1} $[f_{k-1}]$ by x $[y]$ in the definition of π_k $[\phi_k]$ in (8.13)-(8.14). It can again be shown that the weight vectors $\omega_1, ..., \omega_a$ are orthogonal, and that

Proposition 8.3.1.

$$a)\ \omega_{a+1} = \zeta - \Sigma W_a (W_a' \Sigma W_a)^{-1} W_a' \zeta, \quad [W_a = (\omega_1, ..., \omega_a)]. \tag{8.18}$$

b) $\hat{y}_{a,PLS} = \mu_y + \phi_1 \tau_1 + ... + \phi_a \tau_a$ *is of the form* $\hat{y}_{a,PLS} = \mu_y + \beta_{a,PLS}'(x - \mu_x)$ *with*

$$\beta_{a,PLS} = W_a (W_a' \Sigma W_a)^{-1} W_a' \zeta. \tag{8.19}$$

8.4   Theoretical Aspects of Partial Least Squares

It is useful to start by a general representation of the theoretical regression vector, already met in Chapter 3. Let the x covariance matrix Σ have full rank p. Then Σ has a spectral decomposition with positive, different eigenvalues

$$\Sigma = \sum_{j=1}^{q} \lambda_j P_j \tag{8.20}$$

as in Section 3.8.2. Also, as shown there, this implies that the theoretical regression vector $\beta = \Sigma^{-1} \zeta$ can be written

$$\beta = \sum_{j=1}^{q} \gamma_j e_j, \tag{8.21}$$

where there is one unit vector e_j from each of the orthogonal spaces upon which the operators P_j project, and where γ_j are non-negative scalars. Note that the operator P_j projects upon a space of dimension larger than 1 if and only if Σ has coinciding eigenvalues λ_j. Thus q equals p if and only if there are no coinciding eigenvalues.

A further reduction of terms in (8.21) occurs if some of the γ_j's vanish.

Definition 8.4.1. *The eigenvectors e_j occuring in (8.21), that is, those with $\gamma_j \neq 0$ and one for each λ_j are called the relevant eigenvectors. The corresponding principal component scores $e_j'(x - \mu_x)$ are called the relevant components for predicting y.*

With this specification the parameters describing the covariance structure of $x_1, ..., x_p, y$ are Σ and β together with the residual variance σ^2. When p is larger than the number n of observations, and also in some intermediate cases with collinearity, it may be natural to do a model reduction. In these cases it may be adequate to use the word c-variable instead of parameter for Σ, β, σ^2 in agreement with the nomenclature used elsewhere in this book. This use of words is not too important, however.

In any case, the above general concepts turn out to be very useful when we now we go back to PLS. A natural point of departure for a theoretical analysis of PLS is the population algorithm 8.3.1.

Definition 8.4.2. *In general, let S_a be the space spanned by the weight vectors $\omega_1, ..., \omega_a$, and let m be the maximal dimension of this sequence of spaces, i.e., m is the first integer such that (8.18) gives $\omega_{m+1} = 0$.*

Below, I will prove three theorems which characterise S_a and m, and I will show that there is a very close connection between the definitions above. In particular, m turns out to be the number of relevant components.

Theorem 8.4.1 *a) S_a is also spanned by the vectors $\zeta, \Sigma\zeta, \Sigma^2\zeta, ..., \Sigma^{a-1}\zeta$.*
b) m is the least integer a such that $\Sigma^a\zeta$ belongs to S_a.
c) m is the least integer a such that $\beta = \Sigma^{-1}\zeta$ belongs to S_a.

Proof.
A straightforward induction argument is used to prove a), observing that the last term in (8.18) is Σ multiplying a vector which is a linear combination of $\omega_1, ..., \omega_a$. It follows from a) that m is the maximal dimension of the space generated by the vectors of this sequence, so b) follows. Since we have assumed that Σ is invertible, c) is equivalent to b).

Theorem 8.4.2. *a) m is the number of different eigenvalues λ_j such that $e'_j\zeta \neq 0$ for at least one eigenvector corresponding to λ_j.*
b) m is equal to the number of relevant components in x for predicting y.
c) The space S_m is also spanned by the relevant eigenvectors $e_1, ..., e_m$ of (8.21).

Proof.
Using a version of (8.20) including all eigenvectors implies

$$\sum_{k=1}^{a} c_k \Sigma^{k-1} \zeta = \sum_{j=1}^{p} e_j \{ \sum_{k=1}^{a} c_k (\lambda_j)^{k-1} e'_j \zeta \}, \tag{8.22}$$

and this is 0 if and only if

$$\sum_{k=1}^{a} c_k (\lambda_j)^{k-1} = 0 \text{ for all } j \text{ such that } e'_j\zeta \neq 0. \tag{8.23}$$

Let J be the maximal number of such λ_j, and look at the system of equations (8.23) for $a = J$. The determinant corresponding to this set of equations will be a Vandermonde determinant (also called alternant), and this determinant is non-zero if and only if $\lambda_1, ..., \lambda_J$ are different. This implies then that $\zeta, \Sigma\zeta, ..., \Sigma^{J-1}\zeta$ are linearly independent, and that dependence is introduced by adding $\Sigma^J\zeta$ to the set. Thus by Theorem 1 b) we have $J = m$, and a) follows.

In this argument we can also replace $e'_j\zeta$ by $e'_j\beta$, and b) follows then from a) and (8.18). To prove c), we start similarly with

$$\Sigma^{k-1}\beta = \sum_{j=1}^{p}(\lambda_j)^{k-1}e_j(e'_j\beta) \ (k = 1, 2, ..., m)$$

Here we can drop the irrelevant components on the right-hand side, that is, drop the terms with $e'_j\beta = 0$ and by rotation all terms except one when some correspond to the same eigenvalue. Thus by Theorem 3.8.1 a), a basis for S_m can be expressed in terms of the m relevant eigenvectors of Σ, which therefore also must be a basis.

It is a well known fact that the sample PLS algorithm, when carried out to the maximal number of steps p will give the ordinary multiple regression estimator. This fact can be seen by noting that the orthogonal vectors t_k ultimately span the whole $x-$space. The fact has by some been taken as a 'proof' of the consistency of PLS. The following result gives a stronger theoretical result in the same direction: The population PLS algorithm, when carried out in m steps is consistent indeed. In the next section I will discuss a sample estimator which corresponds to a model with m relevant components.

Theorem 8.4.3. *The theoretical PLS regression vector $\beta_{a,PLS}$ will be equal to the theoretical regression vector β if and only if a is equal to the number m of relevant components.*

Proof.
From (8.18) and (8.19) we have

$$\omega_{a+1} = \zeta - \Sigma\beta_{a,PLS}.$$

Hence $\beta_{a,PLS}$ will be equal to $\Sigma^{-1}\zeta = \beta$ if and only if $\omega_{a+1} = 0$.

The above results also give a point of departure for understanding the statistical properties of the ordinary PLS algorithm 8.1.1, since the quantities of this algorithm can be looked upon as estimates of the corresponding parameters given by Algorithm 8.2.1. This point has so far not been much appreciated by the chemometrical literature, and in the statistical literature the general interest around PLS as a general prediction method has been rather limited. Nevertheless some asymptotic calculations under this framework have been published by Helland and Almøy [117].

These asymptotic expressions are relatively complicated. Qualitatively, it turns out that the difference between principal component regression (PCR) and PLS in most cases is relatively small. No method dominates the other. PCR does best when the irrelevant eigenvalues are relatively small or relatively large. PLS does best for intermediate irrelevant eigenvalues. Since the difference is very small for small irrelevant eigenvalues, and since large irrelevant eigenvalues seem to be very rare, this can be interpreted as an, admittedly relatively weak, argument for PLS in this comparison. The conclusions above are confirmed in the systematically designed simulation study by Almøy [11].

Most of the comparisons between regression methods have been done via simulation studies. An extensive discussion of PLS from a statistical point of view, including systematic Monte Carlo simulations, has been given in Frank and Friedman [85]. In these simulations, ridge regression came out best in an overall assessment, followed closely by PLS and PCR, while variable selection did not perform as well as the other methods. The small difference between PLS, PCR and ridge regression were commented upon by the authors by saying that one would not sacrifice much average accuracy over a lifetime by using one of them to the exclusion of the other two. In the discussion, Svante Wold gave arguments to the effect that ridge regression would probably have performed differently under a different simulation design.

In my view, the evaluation of the PLS algorithm should be done in taking into account the relationship to the population PLS model of Section 8.2. The population model with $m < p$ steps is equivalent to a definite restriction of the original model parameter, say, by stating that the population weigh ω_{m+1} at step $m + 1$ is zero. In some sense, it is true that the PLS loadings, weights, etc. give reasonable estimates of the corresponding sample quantities, but these estimates have a very important defect: The sample estimators do not satisfy the restrictions implied by the population model. For instance, the probability that the sample weight w_{m+1} at step $m + 1$ should vanish, given the corresponding population model, is zero.

Thus, any question about finding out in which sense the ordinary PLS algorithm should be optimal, seems to be meaningless. The most we can do is to state the following two questions:

1. In what settings is the model reduction assumed by the population PLS model the most meaningful one?

2. Given the population PLS model with m steps, what are the best possible estimates of the parameters of this model?

A fairly satisfying answer to 1) is given by the theorems above together with the results of Section 3.8. In fact, taking the latter results as a point of departure, we can also to a large extent answer 2).

8.5   The Best Equivariant Predictor

As a summary, in the model with m relevant components we have

$$\beta = \sum_{k=1}^{m} \gamma_k e_k, \tag{8.24}$$

with γ_k being positive constants, and where e_k is an orthonormal set of m vectors. The group G acting upon this vector consists of multiplying each γ_k by a positive constant c_k and rotating each e_k in such a way that the set is still orthonormal.

Lemma 8.5.1. *The group G is transitive on the parameter space for β.*

Proof.
Starting with some β_0 every β of the form (8.24) can be obtained by a transformation of the type described above.

Definition 8.5.1. *Let the loss function under estimation of β be given by*

$$L = (\hat{\beta} - \beta)' S (\hat{\beta} - \beta). \tag{8.25}$$

where S is the non-negatively definite empirical x–covariance matrix.

Lemma 8.5.2. *Whenever the estimator $\hat{\beta}$ is equivariant, the loss function L is invariant under the group G.*

Proof.
Each group element g corresponds to a combination of rotations and scale transformations in $\{\gamma_k\}$, hence to some linear transformation $\beta \to A\beta$, which again must correspond to $x_i \to A^{-1}x_i$ for each data point x_i. This gives $X \to XA^{-1}$ and $S = X'X \to (A')^{-1}SA^{-1}$. For equivariant estimators we have $\hat{\beta} \to A\hat{\beta}$, and L is invariant under the group element g.

The above results are in principle what we need to derive the best equivariant estimator $\hat{\beta}$ under the group G by using Corollary 3.5.5. Strictly speaking this corollary assumes that the group is transitive on the whole parameter space Θ, but using the extension formulated in Theorem 3.5.6, it is enough that the group is transitive on the image space of the function $\theta \to \beta(\theta)$, which is what I proved in Lemma 8.5.1 above.

Recall that PLS, PCR and ridge regression are equivariant under the rotation group. What one in theory could hope to derive now is the best equivariant estimator under the larger group G, using Corollary 3.5.5. This is in a sense exactly what we need, but in this case the invariant posterior is non-proper. This means that the posterior mean, which is the estimator suggested by Corollory 3.5.5, is infinite.

Recently, R. D. Cook et al. [45] have also developed a model reduction technique related to PLS. They use extensively their concept of a central subspace. Their result contains essentially PLS as a special case, and provides some insight into the effectiveness of this method.

See also Section 8.8 below.

8.6 The Case of a Multivariate Dependent Variable

It is well known that the multivariate linear regression model $Y = XB + E$, where Y is an $n \times r$ data matrix and B is a $p \times r$ matrix of regression coefficients, under the maximum likelihood or least squares paradigm leads to nothing new compared to r separate linear regressions. The corresponding fact is not true for PLS. There has been proposed slightly different PLS algorithms for this case [57]. I will not go into detail with this here, but discuss briefly the corresponding model reduction issue.

Let Σ be the $p \times p$ covariance matrix for the x-variables, let B be as above, and let τ be the $r \times r$ covariance matrix. Then the covariance structure of (X, Y) is determined by the parameter (c-variable) (Σ, B, τ). Again the model reduction is determined by a number m of relevant components, but this number must be taken so large that it can cover all r dimensions in Y.

Recall that in the simple case the orthonormal set of eigenvectors e_k in (8.24) can be rotated freely. This is the case also here, but since we now have a matrix B, we must replace the earlier formula by a singular value decomposition of this matrix. Thus

$$B = E\Gamma V' = \sum_{k=1}^{m} \gamma_k e_k v_k'.$$ (8.26)

Here e_k are p-vectors, γ_k are scalars, and v_k are r-vectors. This implies that the group G is composed of scale groups for the γ_k's together with the rotation of the e_k's and rotation in the Y-space.

8.7 The Two Cultures in Statistical Modelling

Breiman [36] initiated an interesting discussion in 2001. Most of his paper is a criticism of the data modelling culture, which he claims is adhered to by 98% of all statisticians. I guess this estimate is approximately true. The common procedure of the data modelling culture is simply what you find in most statistical textbooks, in very many statistical papers and in much statistical practice: Start the statistical analysis by formulating a parametric probability model for the data. Then make all the further discussion relative to this model. This is simply what I called standard statistical inference in Chapter 2.

Breiman's criticism of the standard statistical inference is based on several points, but he mentions three specific items:

First he cites McCullagh and Nelder [152]: 'Data will often point with equal emphasis on several possible models, and it is important that the statistician recognise and accept this.' For instance, in a regression analysis which starts with 30 variables, there are 140 000 five-variable subset models in competition for the best one. There will typically be many models that have residual sum of square (RSS) within 1.0% of the lowest RSS, and several of these models will tell a completely different story.

This point is valid also for some of the alternative model approaches which Breiman promotes, and it can be illustrated by the concept developed in the preceding chapters of the present book: Let the parameters of all competing models together be called a c-variable ϕ, and for model a let the parameter vector be λ^a. Different models may correspond to complementary explanations.

The second item raised by Breiman, is related to Occam's razor: Simpler models are better. A problem, however, is that prediction accuracy and simplicity may be in conflict. Again this is a point also for the alternative model approaches. While the simple tree models (see below) can give poor predictions, the more complicated forests typically gives better prediction.

The third item is related to the curse of dimensionality. Traditionally, the first step in prediction methodology has been to avoid this curse. If there were too many predictor variables, the recipe was to find a few features (functions of the predictor variables) that 'contain most of the information' and then use these features to replace the original variables. Typically one has used variable deletion to reduce the dimensionality. But according to Breiman, recent work has shown that dimensionality can be a blessing.

It is impossible to evaluate Breiman's arguments against a routine use of standard statistical inference without discussing his alternative: Algorithmic modelling. This is a large class of methods containing the following:

1) Classification and regression trees. A predictor $\hat{f}(X)$ is constructed through a series of splits in the multidimensional X-space. The mean in each resulting region is taken as predictor in that region.

2) Random forests grow many classification trees. Each tree gives a classification, and we say that the tree 'votes' for that class. The forest chooses the classification having the most votes.

3) Neural network is a parallel distributed processing network, whose functioning is modeled after the structure of the brain. The output relies on the cooperation of individual neurons within the network. There are variants; the most common one is called the single hidden layer back-propagation network, or single layer perceptron; see [103].

4) Smoothing spline algorithms; in particular Breiman mentions Grace Wahba's research built on theory involving reproducing kernels in Hilbert space [200].

5) Vapnik's support vector machines [198].

This list could have been made much longer. The machine learning community and other groups are constantly developing new algorithms for analysing data, and are improving old algorithms. In particular, the partial least squares algorithms could have been included. With these algorithms we have established a new phenomenon in the preceding sections: It is possible to find a link between the two cultures! The population version of the algorithm is closely related to a specific model reduction in the multinormal model, as shown in Section 8.5.

Let me try on this background to give a personal summary and conclusion of the debate initiated by Breiman. The statistical modelling culture does in fact have many useful applications, not least connected to simple methods like t-tests, F-tests and chisquare tests, or to binomial or Poisson modelling in various biological contexts. Going further, the literature is full of more advanced applications. In the great majority of these cases the distinction between parameter and observation is useful and relevant. What one perhaps can criticise mathematical statisticians for, is that this is taken as the only possible paradigm, and consequently that data analysis is taught at universities as merely a logical deduction from this specific modelling concept. In my view this modelling concept must be supplemented by other tools in order to have a complete and universally useful tool kit for data analysis.

A special problem is of course to understand these other tools. To this end I think it is particularly useful to have a link, as we have in the PLS case to model reduction in a well understood model. Note that it is crucial throughout this argument to keep the distinction between the parameter space and the sample space: Model reduction does in fact boil down just to reducing the parameter space, that is, restricting the set of probability models which is used to 'explain' the data.

8.8 Model Reduction and PLS

The group theoretical discussions above strongly indicate that a statistical understanding of PLS, if possible, should be based upon model reduction.

A very general approach to model reduction in multivariate models have recently been proposed and explored by Cook et al. [46]. Their envelope model is based on the following observation: Every multivariate model contains a matrix M - typically a covariance function - and a space $\mathcal{R}$ - typically the space spanned by the expectation structure. We say that $\mathcal{R}$ is a reducing subspace of M if $M\mathcal{R} \subseteq \mathcal{R}$ and $M\mathcal{R}^{\perp} \subseteq \mathcal{R}^{\perp}$. One can show that $\mathcal{R}$ reduces M if and only if

$$M = Q_{\mathcal{R}}MQ_{\mathcal{R}} + P_{\mathcal{R}}MP_{\mathcal{R}},$$

where $P_{\mathcal{R}}$ is the orthogonal projection upon $\mathcal{R}$, and where $Q_{\mathcal{R}}$ is the orthogonal projection upon $\mathcal{R}^{\perp}$.

Since the intersection of two reducing subspaces of a matrix is itself a reducing subspace, it makes sense to talk about the smallest reducing subspace of the matrix containing a given space. Assume now that M is symmetric and that $\mathcal{R} \subseteq \text{span}(M)$, a weak requirement. Then the following definition makes sense:

Definition 8.8.1. *The M-envelope of $\mathcal{R}$, to be written $\mathcal{E}_M(\mathcal{R})$, is the intersection of all reducing subspaces of M that contain $\mathcal{R}$.*

Note that the column space of M is itself a reducing subspace of M, so this guarantees the existence of the M-envolope. The concept of an M-envelope is thoroughly discussed in [46], and is shown to lead to a useful model reduction in many cases. A discussion of maximum likelihood estimation in the reduced model is also included.

In Helland and Cook [118] it is shown that PLS in its population variant can be derived mathematically from the envelope model. One point of departure is that envelope models can be formulated in many ways. The simplest such model for direct prediction a scalar y from a vextor x is given below. Let first $\Sigma_{xx} = Var(x)$, $\sigma_{xy} = Cov(x, y)$ and $\beta = \Sigma_{xx}^{-1} \sigma_{xy}$. Next, let Γ_1 and Γ_0 be matrices such that $\Gamma_1' \Gamma_0 = 0$, Γ_1 is of rank m and has m columns and $\text{span}(\Gamma_0, \Gamma_1) = \mathbb{R}^p$. Then $\beta \in \text{span}(\Gamma_1)$ if and only if

$$\beta = \Gamma_1 \eta, \tag{8.27}$$

while it follows from [46] that $\text{span}(\Gamma_1)$ reduces Σ_{xx} if and only if

$$\Sigma_{xx} = \Gamma_1 M_1 \Gamma_1' + \Gamma_0 M_0 \Gamma_0', \tag{8.28}$$

where M_0 and M_1 are non-negative definite.

The main result of [118] is the following:

Theorem 8.8.1. *Let x be a centered vector variable of dimension p and let y be a centered scalar variable. Consider the model defined by (8.27) and (8.28). Then this defines a PLS population model with $\leq m$ relevant components. The model is mathematically equivalent to the model in Proposition 8.8.1 below. The model is also an envelope model with $\mathcal{E}_{\Sigma_{xx}}(\text{span}(\beta)) \subseteq \Gamma_1$.*

The theorem is proved from the following result, which is also discussed in Næs and Helland [166].

Proposition 8.8.1. *With the same variables consider the model:*

$$x = Rz + Uv,$$

where R is of rank m and has m columns, $R'U = 0$, $Cov(z, v) = 0$, $Cov(v, y) = 0$ and $\text{span}(R, U) = \mathbb{R}^p$. Then this defines a PLS population model with $\leq m$ relevant components.

This latter result was connected to a discussion of relevant components related to PLS: The regression model has m relevant components if and only if there are exactly m eigenvectors of Σ_{xx} corresponding to different eigenvalues which have non-zero components along the regression vector. This corresponds to m components in the PLS population model. It is very satisfying that these old results now can be coupled to modern general model reduction theory.

Preliminary results indicate what one might expect: Maximum likelihood estimation from the reduced model gives better prediction than sample PLS.

8.9 A Multivariate Example Resembling Quantum Mechanics

A concept which should be central to statistical inference, is that of focusing. Focusing on which parameter to measure was also central to the foundation of quantum mechanics as it was argued for in the previous chapters of this book. This paved the way towards at least some elements of a common foundation. To illustrate this in a different setting, I will end this chapter with a macroscopic example, that is, a statistical example not from the world of quantum mechanics, but containing the elements of learning, choice of measurement and inference at a distance, resembling in this way part of the discussion found in Sections 6.3, 6.4 and 6.10. I am grateful to Harald Martens for the example.

Example 8.8.1.

In Martens and Næs [151] the multivariate regression model with two sets of regression variables T and U was considered:

$$Z = TA + UB + E. \tag{8.29}$$

Here Z is an $n \times k$ matrix of response variables, T ($n \times p$) and U ($n \times q$) may be possible input variables, A ($p \times k$) and B ($q \times k$) are the parameter matrices, and E is the $n \times k$ matrix of errors.

In the discussion in [151], U was an unknown matrix of latent variables. In such cases it is difficult to obtain a good estimate of the parameters A, even if the error E is nearly vanishing.

Also, recall that our definition of a c-variable includes everything that is unknown, so the latent variable U can be included in this c-variable.

For the purpose of illustration, let us neglect the error E, and let us assume as a start that all variables T, U are unknown, so that we can consider the c-variable $\phi = (T, A, U, B)$. It should be clear from the previous chapters that the word 'c-variable' in this book has a wider meaning than what is usually called 'parameter' in statistics, so this example is consistent with that.

Also, assume that we have two distant measurement stations, so that T (and therefore A) are connected to one station, while U (and B) are connected to the other station. Assume that the response Z has been measured and is known, so that one has

$$Z = TA + UB. \tag{8.30}$$

Now assume that we measure T. Let $P_T = T(T'T)^{-1}T'$, and let $V = (I - P_T)U$. Then from (8.30) we know the product

$$VB = (I - P_T)Z. \tag{8.31}$$

But this implies that by using principal component analysis, we can find a considerable portion of the unknown parameter B. (The simplest case is for $q = 1$, when the $k \times k$ matrix $B'B$ is proportional to $Z'(I - P_T)Z$.) This is of course no action at a distance, but information obtained at station 1 implies considerable information about what is unknown at station 2.

On the other hand, $P_T UB = P_T Z - TA$ is unknown since A is unknown and this matrix is premultiplied by T. This must imply that an essential portion of the matrix U is unknown.

From a purely algebraic point of view, there is symmetry between (T, U) and (A, B) in equation (8.30). This means that one in principle can imagine a complementary experiment where T is unknown, but where one gets accurate information about A from some source. By using the same argument again, one now obtains accurate information about a large portion of U, but considerable parts of B is unknown. Again of course, there is no action at a distance.

It is essential here that we have a choice between measuring T or A at station 1, and through this have a choice between getting information on B or on U at station 2. We must imagine that there is some mechanism ensuring that we only are able to measure one of T or A, not both.

By using a slightly more cumbersome argument, we can extend the example to $E \neq 0$, i.e., to non-perfect experiments.

Note that a necessary prerequisite for the argument as given in the example is the restriction that the combination $Z = TA + UB$ is known. The analogue in the EPR experiment discussed earlier is that we know there that we have an equality $\phi_1 = \phi_2$ among the c-variables, or in the ordinary quantum mechanical language we know that we are in a singlet state.

If we accept that an example like this can be understood from the point of view of statistical inference under choice of focus/choice of measurement, we must in my opinion accept that quantum-mechanical phenomena like the EPR experiment can be understood from a similar point of view. And this phenomenon can be understood without assuming that the universe or people's minds are divided into parallel branches. In my opinion the multiple world/multiple minds concepts represent the most abstract and most complicated approach to a theory whose foundation should be concrete and simple.

Chapter 9

QUANTUM MECHANICS AND THE DIVERSITY OF CONCEPTS

9.1 Introduction

The basic concept of a c-variable - a conceptually defined variable - was introduced already in Chapter 1. Typically, such a variable is natural to introduce in contrafactual situations - situations where there are several possible background possibilities, and only one of these possibilities can be realised. In a statistical setting this may mean that one has to choose one out of several possible experiments. The c-variable ϕ can then be taken to be the cartesian product of the parameters of the single experiments. Given the choice a, one must assume that the corresponding experiment is maximal in the sense that the parameter λ^a is a maximal function of ϕ. Several such situations were described in Chapter 2 and in Chapter 7.

Then in Chapter 4 and in Chapter 5 such a situation was taken as a point of departure in my development of quantum mechanics. In Definition 4.6.1. the basic setting was formulated; later 5 specific axioms for this setting were given, and the link to ordinary quantum theory was established in Chapter 5.

One limitation of this approach lies in a specific element of the setting given by Definition 4.6.1: It was assumed there that there is a transitive group G under which the c-variable transforms. Several of the axioms make use of this symmetry group. This means that the theory here is only applicable to physical systems with symmetry. This class of systems is interesting enough: It includes particles with spin, systems of particles with spin and much more, but the existence of a symmetry group is definitively a limitation. One should also recall the recent alternative argument by Wetterich [204] to the effect that quantum mechanics can be derived from classical statistics.

From a mathematical point of view, probability theory and statistical theory are based upon set theory, and then my approach to quantum mechanics is based upon group theory. There is a mathematical theory which generalises both these foundations, namely category theory. Recently, Döring and Isham [64–67] have developed a completely new foundation for physical theories, including quantum mechanics, based on a variant of category theory, namely topos theory. This is motivated from problems in quantum gravity, an area where it is unnatural to talk about an observer, and thus it may seem very different from my own epistemically based theory. However, one should keep in mind that even in statistics, at the very outset an epistemic science, there has been proposed a foundation based on category theory [156]. It may seem like, when using such a general and abstract language, the distinction between epistemology and ontology disappears, or at least becomes less important. The point is that there exists a general conceptual version of quantum theory, and that our more limited theory based on symmetry and focusing in some sense may be looked upon as a special case of this general version.

I will not go into category theory here. On the contrary, very little abstract mathematics will be used in this chapter. But I will use the results of Döring and Isham as an indication that the group theory setting can be generalised. Hence I will drop any reference to symmetry from now on in this chapter, and I will concentrate on the c-variable ϕ and the different choices of focus λ^a. This will lead us into various conceptual contexts where the connection to quantum mechanics may be more or less clear.

Good examples of this are the recent developments by Diederik Aerts and his group in Brussels, who arrive at connections between concepts and quantum theory which are explicit and reasonably obvious. In Aerts at al [6] one finds several examples, in particular one on the state of mind of a person A observing one or two pets. The main point arrived at in this example is that the various states of mind together violate Bell's inequality, the quantum mechanical inequality here discussed earlier in Section 6.3. Later, in Aerts et al [7] the authors remark that the situation considered in the earlier paper introduced conceptual as well as physical contexts, and therefore could not be reduced easily to a purely linguistic situation. A new example was proposed which gave a more clearcut situation. We refer to [7] for more details.

The same research group gave a closer study of modelling concepts and the lattice of contexts in [8], and embedded these entities in a complex Hilbert space, like the one used in quantum mechanics, in [9]. Again, this gives a strong indication of a link between the concepts of the macroworld and the physics of the microworld.

9.2 Daily Life Complementarity

Whenever we aim at completing a reasonably complicated task, we are faced with a multitude of often conflicting goals for this task. As an example, writing a piece of homework as a student, one can aim at minimising the time T used for the homework, or one can try to maximise the quality Q of the homework. Very few people are able to optimise in both respects, at least without a substantial training.

In fact, in many of our daily life decisions we have to decide towards such conflicting goals. We live constantly in such decision conflicts, not least during the interactions with other humans.

Fig. 9.1   A learner-driver.

The goal of this chapter is to argue for a link between the decision process as developed in the earlier chapters in the border area between statistics and quantum theory, over the daily life decision processes to what I feel is also a part of the decision process in psychology, psychiatry and social sciences. Of course, since I am not an insider of any of these sciences, some of my arguments should perhaps be taken with a grain of salt. I have nevertheless taken the chance to include some words of this discussion, partly since it is an opportunity to discuss concepts in a broader setting, and partly since the overall goal of the whole book is to argue for an element of unity in science.

9.3   From Learning Parameter Values to Learning to Make Other Decisions

In modern statistical science it is sometimes said that the aim of a statistical investigation is to use data to learn the value of a parameter. Not least is that the case when the statistical method is an algorithmic one, like in neural networks, but this way of speaking is also used in simple estimation problems.

In any case, the statistical learning situation requires first a conceptually defined context, in order to describe the whole setting and the model used for estimation/learning. Then there are data, and there is a focus parameter λ^a that we want to estimate. All the model parameters, together with the parameters in the context may be called the c-variable ϕ. Finally, there may be a learning criterion, like maximum likelihood, and there may be a final criterion by which the estimator can be judged, like bias together with variance.

The learning in psychology is similar, but somewhat more involved. Think of the task of learning a skill, to be concrete, say, learning to drive a car. There is first a conceptually defined context, consisting of a description of the basic functioning of the car, of its engine and of the instruments that one needs to know about, of the trafic rules and so on. Then in any driving situation there are data in the form of visual impressions from the road, from other cars and so on. Given a situation a, including the driver's intention in that situation, the driver wants to use the data to make an optimal decision $\hat{\lambda}^a$. This includes taking actions with the steering wheel, the gear, the gas pedal and so on. The driver has certain criteria in the situation to help him in making the best possible decisions. To begin with, these criteria are attempted in a fairly conscious and perhaps somewhat clumsy way. Then it takes a certain skill in order that the decision process shall be really automatic and at the same time good according to some objective criterion.

9.4   Basic Learning: With and Without a Teacher

Imagine again the situation of the learner-driver. There are really many situations a that he has to learn to tackle in a relatively short time. Some of these situations will turn up during his driving lessons. Then he will be able to try out what he thinks is the best decision. Say that he meets a set of situations $\mathcal{A}_0$ during his driving lessons; some of these situations may be repeated. Each time he meets a situation $a \in \mathcal{A}_0$ he makes what he thinks is the best possible decision. He will perhaps get some feedback from the reaction of the car, from other drivers and finally from his teacher. All this may make him take a better decision next time a occurs.

The teacher is of course an important factor in this. In theory it is possible to learn to drive a car without a teacher, but, even where this is legal, it takes a long time. One function of the teacher is to argue for the objectively best decision in each situation a. This is theory, but if the pupil is just learning to drive, and if the teacher is experienced, it may not be far from the truth. Then given some situation a, the teacher, if he in addition is a good pedagogue, will, after the pupil has made his decision, be able to give him valuable feedback on how the decision really should be. In practice the teacher will also have his bad habits, or he may not be clever enough to transfer his knowledge to the pupil. This may make the resulting education less valuable than it could have been.

In any case, after the pupil has finished his education, he will continue driving, and ultimately he will meet all relevant situations $a \in \mathcal{A}$. Each time he meets a new situation, he has to try out a new decision. This may not be the best one the first time a new situation is met, but he will have the advantage of earlier experience, and may, if the experience is used correctly, become a better and better driver. A part of this latter learning process will be to use the conceptual setting given by ϕ to transfer skill used earlier to the process of making good decisions in new situations a.

9.5 On Psychology

We are all good or bad teachers in one way or another. One important role that many of us have or have had, is that of a parent. Some years ago, certain psychologists, partly influenced by aspects of Freudian theory, had a tendency to blame the child's parents when a child developed a psychiatric illness. Today one usually look for a diversity of possible causes. However, the following example from Bateson [20], describing the phenomenon of double bind, illustrate a kind of parent behaviour which definitively may be detrimental.

According to [20] the family situation of a schizophrenic may have the following characteristics:

A child whose mother becomes anxious and withdraws if the child responds to her as a loving mother. That is, the child's very existence has a special meaning to the mother which arouses her anxiety when she is in danger of intimate contact with the child. At the same time the mother looks upon these feelings as forbidden.

In this situation the mother will be expressing at least two different messages: λ^a: hostile or withdrawing behavior whenever the child approaches her; and λ^b: simulated loving or approaching behavior which is aroused when the child responds to her hostile and withdrawing behavior, as a way of denying that she is withdrawing.

In a given situation, both these messages may both be intended and perceived through the simple words: 'Go to bed, you're very tired and I want you to get your sleep'. While the child might respond literally to this verbal message, it is impossible for him to react adequately at the same time to both the perceived messages λ^a and λ^b. The point from a psychiatric point of view is that a child which repeatedly experiences such double messages, will get his feelings confused, and may later develop difficulties in his relationship to other humans.

Let this small example illustrate one type of insights from modern psychiatry. There are many schools within psychology and psychiatric therapy; most of them are in one way or other influenced by the writings of Sigmund Freud. Freud himself got much of his insight from introspection, and in modern psychology much is learned through case studies and conceptual discussions. This may be called the qualitative part of psychology.

On the other hand, some psychiatrists feel that knowledge obtained in this way may be uncertain and partly subjective. There is a certain tendency today to seek objective, empirical knowledge in psychiatry to a larger degree. A recent example is Høglend et al [128], a fully empirical study of what psychiatrists call transference interpretation in therapy: The therapist takes a role where he and the patient get the chance to play out threumatic feelings during the therapy hour. This has been a debatable mechanism in psychotherapy, but according to Freud's theory it should be useful for patients with good resources. The present investigation was the first empirical study ever of transference interpretation. It involved 7 therapists who had been given a 4 year training in making a deliberate choice between transference interpretation and other treatments. Then 100 patients were given a blind random choice of treatment during a year and finally followed up in a 2 year period. The statistical analysis used a longitudinal mixed model, and the clear result was contrary to what Freud had told us: Patients with poor resources were the ones who benefitted most from transference interpretation.

Having been a statistical consultant of this study, I also experienced what I have argued for elsewhere in this book: Prior to the analysis a meaningful discussion on the choice of statistical model using a conceptual framework ϕ. We ended up with consensus on a model with total parameter λ, including one covariate. One outlier case was deleted prior to the final analysis.

I also observed elements of cultural differences, the most clearcut being: In this community it was a tendency to test statistical hypotheses at the 10% level in order to get increased power, a policy that I have never seen elsewhere.

In conclusion, psychology is a science which develops in many ways: Through theoretical discussions, through introspections, through case studies and through statistical investigations. In particular, statistics has an increasingly important role to play here. But it can in no way function alone. In order that statistics as a tool shall do what it is supposed to do in psychological sciences, it is important that it functions in a conceptual framework related to the same sciences.

9.6 On Social Sciences

The situation is similar in sociology and other social sciences. Here the natural framework for discussion is the concept of positivism developed by August Comte [44] in the middle of the 19th century. Positivism is a philosophy that states that the only authentic knowledge is scientific knowledge, and that such knowledge can only come from positive affirmation of theories through strict scientific methods. Today, many will add that scientific methods should be taken as empirical, statistical methods, in sociology often specialised to opinion polls and similar techniques.

Even though much of the discussion in sociology today is methodological, in this section I will go back to the fundamental concepts of Comte. In the last couple of decades there has been an increasing critique of positivism among certain groups of sociologists: Man is simply too complicated to be fully described by the objective terms used in natural science, it is said.

One of the first of these critics was Skjervheim [187]. Skjervheim was a philosopher, and he relates his discussion to the antagonistic philosophical currents that exist today: The naturalistic-positivistic-pragmatic trend and the phenomenological-existentialistic trend. One can even find a geographical distribution of these trends: In the Anglo-American world as well as in Scandinavia the former trend has been dominating, while in Germany, France and the Spanish-speaking world the latter is dominating.

Very simply put, one might say that the present book contains elements of both these traditions: From a conceptual framework ϕ one *chooses* a maximal question λ, and then investigates this question through objective naturalistic methods.

A related theme, discussed extensively by Skjervheim in his positivism critique, is that of a truthfunctional language and the implied thesis of extensionality. The ideal language used by Russell in his Principia Mathematica was *truthfunctional*: It is constructed in such a way that the truth of a complex proposition is only dependent upon the *truth* of its constituent propositions, not their meaning. An *extensional* language means the following: If two propositions a and b have the same truth-value, then we can substitute a for b in any context were the latter occurs (and vice versa) without thereby changing the truthvalue of the context.

Skjervheim's opinion is that an extensive truthfunctional language might well be used in natural science, but such a language is too poor to be used in any science of Man. This kind of language will exclude all 1) modal statements ('It is necessary that p'); 2) semantical statements ('The words "p" mean p') and 3) belief-sentences ('A believes p'). Several examples are given to justify a claim that behavior cannot be adequately described in an extensional language for the purpose of social science.

The limitation of language is also an aspect of the last part of his critique. But more broadly, it is related to the theme of intersubjectivity: Another person may in some connections be an object that I may gain certain experiences about. In other connections he may be a subject that I can share experiences with. In particular, the social scientist, when writing, cannot escape this situation: Those about whom he writes are partly identical to those for whom he writes.

Consider the following statements which may be uttered by a seminar holder in some discipline:

'When I gave this seminar last week, everybody understood what I have just said.'

'I hope that you understood what I have just said.'

'I am sure that you understood what I have just said.'

The first statement is of a truth-functional type, while the two last express various degrees of belief. The point is that in the given situation he could never express a statement of the first truth-functional type about the understanding today. Thus the context limits what it is possible to say.

Another point is that the three statements can be taken as answers to three differently focused questions, related to the type of questions that have been treated earlier in this book.

In these sentences, I have only very briefly discussed a few aspects of a very big and complicated debate: That of positivism in social sciences and in other sciences on human beings. The relation to my interpretation of quantum mechanics is also only touched upon.

But it is amusing that Skjervheim [187], when he wants to give an example where the objective, naturalist way of thinking in science is dominating, repeatedly mentions physics. Quantum mechanics was of course well developed in 1959, but it was not perhaps too well known to that author.

But, after all, perhaps the main difficulty that many physicists have seen in quantum mechanics, has been that it has turned out to be very difficult to fit into a completely objectivistic way of thinking. And perhaps it shouldn't; it may fit better in with other sciences in that way.

Chapter 10

EPILOGUE

For a moment, go back to the learner-driver situation of Chapter 9. One may discuss whether or not the same mechanisms can be transferred to the task of learning science. In my view there are both differences and parallel elements here. In science there is a larger body of theory to learn; hence rational and concious elements take a bigger place. On the other hand, in learning science there is also an important element of learning a skill. Students do their homework, make their exercises in classes and take their exams, and get feedback from teachers on all this. PhD students and young faculty give talks on international meetings and get feedback from the audience. Older faculty learn from discussions with colleagues.

There is one more difference from the learner-driver situation: All our teachers belong to the same science. Of course there are many differences in detail between the teachers, but their basic way of thinking is the same. This has many advantages: First, it is much easier to communicate when we share the same basic concepts. In fact, the common platform may be seen as a prerequisite for the impressive progress that science has made in many areas.

But at the same time: The common way of thinking within each science may have made communication across sciences difficult.

In particular, there is a common particular way of thinking in theoretical statistics, concentrating only on single experiments of the standard type with a model constructed from a parametric class of probability distributions. Also, there is a common particular way of thinking in quantum theory, focusing exclusively upon state vectors as unit vectors in a complex Hilbert space and upon comparisons with classical mechanics. These separate bases have turned out to be very useful in the development of each science. It is wrong to say that the separation between these two areas are completely culturally determined. It also plays a role that the mathematics of each field has a separate deep aestethic appeal. And of course, there are basic differences between the physics of the microworld and the problems associated with gaining information from macroscopic data.

Nevertheless, I will claim that there are some cultural differences, and in this book I have tried to sketch the elements of a common basis. It may still be useful to think separately when contemplating questions for which the separate conceptual basis is best suited. But my hope is that I have argued sufficiently for the common basis that others will be convinced of its usefulness, and perhaps also that other people may help in developing it further. In my opinion it can already be used on questions which was difficult to address before. For this I refer to the discussion in earlier chapters.

Some readers may be sceptical to the thesis that separate sciences have separate cultural development to the extent that this may be of importance. To counter this scepticism, it may be enough to point at all the other areas in the world where separate learning has lead to separate cultural development. Some of this is connected to religion, some to ethnic group and some to geographical region. Children have been raised in different ways, pupils have been given different education, and adults have developed different senses of belonging. One should not study this world much before realising that cultural differences are extremely strong forces; why shouldn't science also be affected by it?

Another argument may be related to the development within the sciences themselves. The international statistical community thinks differently today than 30 - not to mention 50 - years ago. Part of this is due to the technological development, other parts to new theoretical achievements. But in addition to this: If we regard some of the opinions held by statisticians 50 years ago as out of fashion, don't we at the same time admit that there is a fashion component in science, and perhaps even also that this may be of relevance today?

The culture in science may be related to which questions are being focused upon, both in terms of what is being discussed, or in terms of what research questions that may be regarded as interesting. More generally one might say that the culture in some way or other determines important parts of the context of the various research questions.

More precisely, Ralph D. Stacey has put the following definition forth in [191]: 'Culture is a set of attitudes, opinions and convictions that a group of people share, about how one should behave against each other, how things shall be evaluated and done, what questions that are important and which answers that are acceptable. The most important elements of culture are subconscious and cannot be enforced from the outside.'

Of course, cultural elements can be positive, but cultural clashes between different groups can have very negative consequences. To state things in its extreme form: In a world constantly threatened by war and terror partly on the basis of cultural conflicts, one could hope that there is at least some elements of the society which seeks just objective values across all cultural differences. One hope is that in the long run, parts of science will, in addition to all its other positive values, concentrate on being such an element of society. A prerequisite for this is that we at least to a certain extent, and related to these parts of science, develop a common language, so that we can communicate across the scientific professions.

So the final question is: Is it possible to translate into one common language, and thus create a mutual understanding between cultures of empirical science? My claim is that this book contains many indications in that direction. Others have had similar thoughts before; for instance the physicist John A. Wheeler may have thought of this when he said: 'Progress in science owes more to the clash of ideas than to the steady accumulation of facts'. Admittedly, this citation can be interpreted in many directions, but in my opinion it is potentially very useful to bring ideas together that have originated in completely different fields.

Put quite simply, science implies searching for the truth. In particular then it implies searching for good questions and then for convincing answers to these questions. And when one searches for something, it can quite generally be wise to look several places.

Similarly, to have a broad enough background for one's own opinions, it can be useful to have had contact with several cultures.

But of course the whole process of searching for truth in science is difficult. This is the first point that can be interpreted from Albert Einstein's famous saying 'Subtle is the Lord, but not malicious'. The last, optimistic point is that, after all, the search should not be impossible.

Our point of departure in this search is imperfect; one general hint at this may be found by citing 1. Corintheans 13.9: 'For we know in part.' This brings us briefly into the field of religion, a field that I have great respect for. But to me it is a difficult point that our beliefs about the final things tend to depend upon whether we are born in Rome, in Western Norway or in Teheran, or in general which people you are influenced by. Again, the cultural differences point at a serious stumbling block. Of course, the worst effect of cultural differences has shown itself in regional armed conflicts, on Balkan, in the Middle East and many other places. In these cases international newspapers are full of contributions supporting one party or the other. Relatively few people seem to realise that it is the cultural conflict which is the problem. In most cases this conflict is so serious and so complicated that it is impossible to take the simplified view of saying that one of two or more parties is right or wrong. Realising this may help in focusing on the real problems.

Modern society and modern science are faced with very many challenges: We have the struggle for a better global environment, the fight against poverty, the search for better medicines, for a deeper understanding in genetics, of the universe and of the basic building blocks of matter. One cannot of course expect a united science over all these fields, but what on can hope for, is a more unified understanding of the process of gaining information from observations. In this process, mathematics will certainly play an important part, but one should also be willing to include the experience gained in the various empirical fields; in fact, this can to some extent be used as a calibration on how to use the mathematics.

Ultimately it should not be impossible to head in the direction of something like a common basis for understanding across various professional scientific groups. This does of course not mean that we all should have the same opinions about science, but there should be some basic context that we were able to agree on. And what one could hope for, was that this conceptual context should be common for many different professions.

If one somehow could see a development where a type of multicultural understanding is first indicated in certain specific parts of science, then one could at least hope for *some* inspiration to other areas of human endeavor, too. Since science - as represented by physics and by data analysis - is probably the part of human endeavor which is most exact, one can argue that its most natural to start by seeking for objectivity here.

I know that parts of this book are controversial, and I am willing to discuss any questions that can be raised. I am also willing to modify details, if necessary. However, the arguments shall be very strong indeed before I am willing to give up the final aims of the search, as indicated in the last paragraphs above. Simply put, I feel that a closer unification of scientific methods is useful - and possible.

APPENDIX

A.1 Mathematical Aspects of Basic Statistics

A.1.1 *Kolmogorov's Axioms*

Kolmogorov's axioms for probability theory are derived from set theory. The basic probability space, the certain set, is denoted by Ω, and the events for which probabilities can be specified, are subsets of Ω. Specifically, these events constitute a σ-algebra $\mathcal{F}$, a concept defined by the axioms:

1. The certain set Ω is in $\mathcal{F}$.
2. If $A \in \mathcal{F}$, then its complement $A^c = \{\omega \in \Omega : \omega \notin A\}$ also is in $\mathcal{F}$.
3. If the events $A_1, A_2, \ldots$ are in $\mathcal{F}$, then their union $\bigcup_{i=1}^{\infty} A_i$ also belongs to $\mathcal{F}$.

The probability P is assumed to be a normed measure on these events, which in axiomatic terms means that it satisfies:

1. For every A we have $P(A) \geq 0$.
2. If the events $A_1, A_2, \ldots$ are in $\mathcal{F}$, and if pairs of events are disjoint, i.e., $A_i \cap A_j = \emptyset$ for $i \neq j$, then $P(\bigcup_{i=1}^{\infty} A_i) = \sum_{i=1}^{\infty} P(A_i)$.
3. $P(\Omega) = 1$.

From these axioms many properties of probability models may be derived. Two simple consequences are that $P(\emptyset) = 0$ and $P(A^c) = 1 - P(A)$. With some more effort one can derive a general formula for the probability of a union of events, simplest in the case of two events:

$$P(A \cup B) = P(A) + P(B) - P(A \cap B). \tag{A.1}$$

A.1.2 *The Derivation of the Binomial Distribution*

Assume an experiment which satisfies the three assumptions of Example 1.8.1. Then each such experiment will result in a sequence of n outcomes of the form

$AABA\dots B$, where $B = A^c$. By the constant probability assumption 2, the independence assumption 3 and the definition of independence (1.4), each such sequence which contains exactly s A's will have a probability of $\theta^s(1-\theta)^{(n-s)}$. By simple combinatorics there are $\binom{n}{s}$ such sequences. An application of the second of Kolmogorov's axioms finishes the proof of the derivation of the binomial distribution.

The expectation and the variance of this distribution can be found from the frequency function, but the following argument is simpler: From the definition of the binomial variable, we can write $y = i_1 + \dots + i_n$, where the indicator variables i_j are independent with $i_j = 1$ if the j'th voter answers A, otherwise $i_j = 0$, hence $E(i_j) = 0 \cdot P(i_j = 0) + 1 \cdot P(i_j = 1) = P(i_j = 1) = \theta$, with a similar way to calculate the variance. Since the expectation operator E is linear, and since $Var(y) = E(y-\mu)^2 = E(y^2) - \mu^2$, it is then a fairly straightforward exercise (using independence of the i_j) to show that $\mu = E(y) = n\theta$ and $Var(y) = n\theta(1-\theta)$.

A.1.3   *The Normal Distribution and Series of Observations*

The *characteristic function* $\phi(u) = E(\exp(iuy))$ for a random variable y is a useful concept. It characterises the distribution of y in the sense that two variables with equal characteristic function also must have the same distribution. Also, for sets of independent variables the characteristic function of their sum is the product of the individual characteristic functions.

By a direct integration we find that the characteristic function for the normal (μ, σ^2) distribution is

$$\phi(u) = \exp(iu\mu - \frac{\sigma^2 u^2}{2}). \tag{A.2}$$

From this we see first that $a + by$ is normal $(a + b\mu, b^2\sigma^2)$ when y is normal (μ, σ^2). Furthermore, for a set of normal independent variables $y_1, \dots, y_n$ we have that $a + b_1 y_1 + \dots + b_n y_n$ is normal with the appropriate expectation and variance. In particular, for a normal measurement series (a set of independent identically distributed normal variables) the mean $\bar{y}$ is normal $(\mu, \sigma^2/n)$.

Generalizing to the multivariate case, one can show by characteristic functions that each $y_i - \bar{y}$ is independent of $\bar{y}$ for such a measurement series, hence that $\bar{y}$ and s^2 (Definition 2.3.2) are independent.

Take now for simplicity $\mu = 0$ in the measurement series; the general case is similar. Since

$$s^2 = \frac{1}{n-1} \sum_{i=1}^n (y_i - \bar{y})^2 = \frac{1}{n-1} \sum_{i=1}^n y_i^2 - n\bar{y}^2, \tag{A.3}$$

we have that

$$\frac{(n-1)s^2}{\sigma^2} + \frac{n\bar{y}^2}{\sigma^2} = \frac{1}{\sigma^2} \sum_{i=1}^n y_i^2. \tag{A.4}$$

By the definition in Section 2.5, the right-hand side of (A.4) has a chisquare distribution with n degrees of freedom. Since $n\bar{y}^2/\sigma^2$ trivially is chisquared with one degree of freedom, this strongly suggests - and in fact it can be proved rigorously by using characteristic functions - that

$$\frac{(n-1)s^2}{\sigma^2} \tag{A.5}$$

has a chisquare distribution with $n-1$ degrees of freedom.

A.1.4 *Some Results for Linear Models*

The multivariate normal distribution (μ, Σ), where μ is a q-vector and Σ is a positive definite $q \times q$-matrix can be uniquely defined as the distribution of a vector z such that $a'z$ is normal $(a'\mu, a'\Sigma a)$ for each a. This can be shown by using appropriate characteristic functions.

Maximum likelihood in the linear model $y = X\beta + e$ (e multivariate normal $(0, \sigma^2 I)$), where X is a $n \times p$ matrix, can be shown to be equivalent to least squares, i.e., minimising

$$\| y - X\beta \|^2 = (y - X\beta)'(y - X\beta) = y'y - 2\beta'X'y + \beta'X'X\beta. \tag{A.6}$$

Differentiation with respect to β here leads to the normal equations $X'X\beta = X'y$, hence $\hat{\beta} = (X'X)^{-1}X'y$. Using the model, the properties of this vector estimator are found as follows:

$$E(\hat{\beta}) = (X'X)^{-1}X'E(y) = (X'X)^{-1}X'X\beta = \beta, \tag{A.7}$$

$$V(\hat{\beta}) = V((X'X)^{-1}X'e) = (X'X)^{-1}X'V(e)X(X'X)^{-1} = (X'X)^{-1}\sigma^2, \tag{A.8}$$

where $V()$ denotes the covariance matrix.

It follows from linearity that $\hat{\beta}$ is multivariate normal $(\beta, (X'X)^{-1}\sigma^2)$, that $X\hat{\beta}$ is multivariate normal $(X\beta, \sigma^2 P)$ with the projection operator $P = X(X'X)^{-1}X'$ and that $y - X\hat{\beta}$ is multivariate normal $(0, \sigma^2 Q)$ with the projection operator $Q = I - P$.

From these properties one can show that $SSR' = (\hat{\beta} - \beta)'X'X(\hat{\beta} - \beta)/\sigma^2$ has a chisquare distribution with $p + 1$ degrees of freedom, that $SSE' = (y - X\hat{\beta})'(y - X\hat{\beta})/\sigma^2$ has a chisquare distribution with $n - p - 1$ degrees of freedom, and that these two sums of squares are independent.

A slight modification of this result (using the fact that the first column of X is 1 or at least, the vector 1 belongs to the span of the columns of X) can be used to prove that the test variable (2.50) has the required F-distribution under the null hypothesis. The same results combined with the normality of $\hat{\beta}$, and the independence of $\hat{\beta}$ and s^2, which can be shown to hold here, gives the t-distribution of the testvariables for single β-components. Details are given in textbooks such as [177].

A.1.5   *On the Fisher Information*

Consider a standard statistical model with parameter θ, here assumed one-dimensional, and with a probability density q^θ for the observations y. For fixed y let

$$l(\theta) = \ln L(\theta) = \ln q^\theta(y). \tag{A.9}$$

Since $l(\theta)$ depends on y, it will be a random variable.

Regularity conditions needed for the arguments below are given for instance in [27], p. 179-180.

First, define the score by $s(\theta) = \frac{\partial}{\partial \theta} l(\theta)$ and the Fisher information by

$$I(\theta) = E^\theta(s(\theta)^2) = \int (\frac{\partial}{\partial \theta} \ln q^\theta(x))^2 q^\theta(x) dx. \tag{A.10}$$

Since $\int q^\theta(x) dx = 1$, we have $0 = \frac{\partial}{\partial \theta} \int q^\theta(x) dx = \int (\frac{\partial}{\partial \theta} \ln q^\theta(x)) q^\theta(x) dx = E^\theta(s(\theta))$. Hence $I(\theta) = Var(s(\theta))$. By a second differentiation one can also show that $I(\theta) = -E^\theta(\frac{\partial^2}{\partial \theta^2} l(\theta))$.

We now prove the Cramer-Rao Theorem, which says: Let $\hat\theta$ be any unbiased estimator of θ. Then

$$Var^\theta(\hat\theta) \geq \frac{1}{I(\theta)}. \tag{A.11}$$

Proof.

By differentiation of

$$\theta = E^\theta(\hat\theta) = \int \hat\theta(x) q^\theta(x) dx,$$

we find

$$1 = \int \hat\theta(x)(\frac{\partial}{\partial \theta} l(\theta)) q^\theta(x) dx.$$

Using the Cauchy-Schwarz inequality on this together with the identity

$$0 = \theta E^\theta(s(\theta)) = \int \theta(\frac{\partial}{\partial \theta} l(\theta)) q^\theta(x) dx,$$

(A.11) follows.

Multivariate and other generalizations of these results are given in [27].

A.1.6   *Prediction Errors in Example 2.15.1.*

From the model $y_i = \beta_0 + \beta x_i + e_i$ (errors e_i independent and normal $(0, \sigma^2)$) the prediction error at a point x_0 with $\mu_0 = \beta_0 + \beta x_0$ is found from the difference between a hypothetical observation y_0 at that point and the corresponding predicted observation from the model $\hat y_0$:

$$\begin{aligned} P = E(y_0 - \hat y_0)^2 &= \sigma^2 + E(\mu_0 - \hat y_0)^2 \\ &= \sigma^2 + Var(\hat\beta_0 + \hat\beta x_0) = \sigma^2 + \frac{\sigma^2}{n} + x_0^2 \frac{\sigma^2}{\sum x_i^2}, \end{aligned} \tag{A.12}$$

where the last expression is found from the fact that $\hat{\beta}_0 = \bar{y}$ and is uncorrelated with $\hat{\beta}$ in this case; this and $Var(\hat{\beta})$ is found from (A.8).

From the reduced model corresponding to putting β equal to 0 in the model the prediction error is

$$P^R = E(y_0 - \bar{y})^2 = \sigma^2 + E(\mu_0 - \bar{y})^2$$
$$= \sigma^2 + Var(\bar{y}) + (E\bar{y} - \mu_0)^2 = \sigma^2 + \frac{\sigma^2}{n} + \beta^2 x_0^2, \tag{A.13}$$

since $E(\bar{y}) = \beta_0 + \beta\bar{x} = \beta_0$ since $\bar{x} = 0$ and $\mu_0 = \beta_0 + \beta x_0$.

A.2 Transformation Groups and Group Representation

A.2.1 *Further Properties of Group Actions*

Adding a group to a statistical model specification is often of interest, and does have consequences, see Lehmann and Casella [146]. First let a group G act on a measurable sample space S. Measurability questions are ignored here, as is common when discussing transformation groups; a full account of this aspect is given in Varadarajan [199]. Basic topological assumptions are made throughout the book, though: Both G, S and the parameter space Θ are assumed to be locally compact Hausdorff such that each point has a compact neighborhood and the topology has a countable base (cf. [68]).

Orbits are defined in Section 3.1.1, while group actions on the parameter space and on the sample space in statistical models are introduced in Section 3.2. Under conditions as given below, each set of orbits can be given an index. The orbit index in the sample space will always have a distribution which depends only upon the orbit index in the parameter space (Lemma 3.2.1).

Concentrate now on the group G acting on the total parameter space Φ. Similar concepts can be defined for the other group actions discussed above. The group G is also assumed to have a topology.

We assume, as is commonly done, that the group operations $(g_1, g_2) \mapsto g_1 g_2$, $(g_1, g_2) \mapsto g_2 g_1$ and $g \mapsto g^{-1}$ are continuous. Furthermore, we will assume that the action $(g, \phi) \mapsto \phi g$ is continuous for $\phi \in \Phi$. An additional condition, discussed in Wijsman [206], is that every inverse image of compact sets under the function $(g, \phi) \mapsto (\phi g, \phi)$ should be compact. A continuous action by a group G on a space Φ satisfying this condition is called *proper*. This technical condition turns out to have useful properties and will be assumed throughout this paper. When the group action is proper, the orbits of the group can be proved to be closed sets relative to the topology of Φ.

For fixed $\phi \in \Phi$, a *stabilizor* H of G is defined as $\{h : \phi h = \phi\}$. These are transformed within orbits of G as $H \mapsto g^{-1} H g$.

A.2.2   *Haar Measure and the Modular Function*

Since in the literature ([158, 121]) Haar measures are introduced as Radon measures, that is, measures on functions rather than sets, we take this as a point of departure. Let G be a locally compact group, and let C_G be the space of continuous functions on G with compact support. Then one can define in a unique way (except for a multiplicative constant) two positive Radon measures, a left Haar measure satisfying

$$\mu_G(f(g\cdot)) = \mu_G(f(\cdot)), \qquad (A.14)$$

and a right Haar measure satisfying

$$\nu_G(f(\cdot g)) = \nu_G(f(\cdot)) \qquad (A.15)$$

for all bounded continuous functions f on G and for all $g \in G$. By the Riesz theorem, ordinary measures are associated with these through

$$\mu_G(f) = \int_G f(g)\mu_G(dg) \text{ and } \nu_G(f) = \int_G f(g)\nu_G(dg). \qquad (A.16)$$

Definition A.1. *(A.16) defines the left and right Haar measures on the group.*

These left and right Haar measures satisfy

$$\mu_G(gD) = \mu_G(D) \text{ and } \nu_G(Dg) = \nu_G(D) \text{ for } D \subset G. \qquad (A.17)$$

Again, μ_G and ν_G are uniquely defined up to a multiplicative constant.

If G is compact or commutative, or if it is finite or countable, we can take $\mu_G = \nu_G$. In the compact case, the Haar measure can be taken to be a probability measure, otherwise, it is an unnormalised measure.

Introduce temporally for some fixed $g \in G$ a new measure by $\mu_{g,G}(D) = \mu_G(Dg))$ for all D. Then

$$\mu_{g,G}(hD) = \mu_G(hDg)) = \mu_G(Dg) = \mu_{g,G}(D).$$

Hence $\mu_{g,G}$ is left invariant, and by the uniqueness of such measures we must have

$$\mu_{g,G}(D) = \Delta_G(g)\mu_G(D)$$

for some scalar function Δ_G.

Definition A2. *The function Δ_G is called the* modular function *of the group* G.

Lemma A1. *The modular function is a* homomorphism*:*

$$\Delta_G(gh) = \Delta_G(g)\Delta_G(h). \qquad (A.18)$$

Proof.

$$\Delta_G(gh)\mu_G(D) = \mu_{gh,G}(D) = \mu_G(Dgh) = \Delta_G(g)\Delta_G(h)\mu_G(D),$$

and (A.18) follows.

Furthermore,

$$\mu_G(Dg^{-1}) = \mu_{g^{-1},G}(D) = \Delta_G(g^{-1})\mu_G(D) = (\Delta_G(g))^{-1}\mu_G(D), \qquad (A.19)$$

since $(\Delta_G(g))^{-1} = \Delta_G(g^{-1})$ follows from the homomorphism. This implies that $\lambda(dg) = \Delta_G(g^{-1})\mu_G(dg)$ is right invariant, since for fixed h one has:

$$\lambda(dgh) = (\Delta_G(gh))^{-1}\mu_G(dgh) = \Delta_G(g^{-1})\Delta_G(h^{-1})\Delta_G(h)\mu_G(dg) = \lambda(dg), \quad (A.20)$$

so:

Lemma A2.

$$\mu_G(dg) \propto \Delta_G(g)\nu_G(dg). \qquad (A.21)$$

Finally, by a similar argument we show that $\Delta_G(g^{-1})\nu_G(dg^{-1})$ is a right invariant measure, which then must be proportional to $\nu_G(dg)$. Considering the special case $g = e$, we see that the constant of proportionality must be 1, so:

Lemma A3.

$$\nu_G(dg^{-1}) = \Delta_G(g)\nu_G(dg). \qquad (A.22)$$

So far we have only talked about measures on the group itself. A measure ν on the space Θ (or on any space on which a group G acts) is called relatively invariant with multiplier $\delta(\cdot)$ if $\nu(Eg) = \delta(g)\nu(E)$ for all g and E.

Regarding relative invariant and in particular right invariant measures on the parameter space, the following result due to Weil is proved in Wijsman [206], Theorem 2.3.13(c):

Theorem A1. *Consider an orbit in* Θ *given by* $\Psi = \{\theta = \theta_0 g; g \in G\}$ *for some* θ_0*. Let the stabilizor H be defined by* $H = \{h : \theta_0 h = \theta_0\}$*. Then there exists a relatively invariant measure (see (3.2)) on* Ψ *with multiplier* $\delta(\cdot)$ *if and only if*

$$\Delta_H(h) = \delta(h)$$

for all $h \in H$. This measure is unique except for a strictly positive constant factor of proportionality, and can be defined by $\nu_G(\beta^{-1}(E))$*, where ν_G is a corresponding relative invariant measure on G, defined as in (3.2), and* $\beta : G \mapsto \Psi$ *is defined by* $\beta(g) = \theta_0 g$*.*

As a special case, a measure which is invariant as defined in Section 3.3, and induced by right invariant Haar measure on G, exists if and only if $\Delta_H(h) = 1$ for all $h \in H$. In particular this holds if H is compact, which is the case if the action of G on Θ is proper and the group is locally compact.

A.2.3   Proofs Concerning Orbits, Model Reduction and Estimation of Orbit Indices

Proof of Theorem 3.5.6.

It follows from equation (3.6), summed over the orbits in the sample space and from Theorem 3.5.2 that $\hat{\eta}_\tau$ minimises the risk in orbit τ corresponding to the quadratic loss function.

To prove that $\hat{\eta}_\tau(y)$ is equivariant under G, fix some group element h. Then

$$\hat{\eta}_\tau(yh) = \frac{\int_G \eta(\theta g) q^{\theta g}(yh) \mu_G(dg)}{\int_G q^{\theta g}(yh) \mu_G(dg)}$$

$$= \frac{\int_G \eta(\theta g) q^{\theta g h^{-1}}(y) \mu_G(dg)}{\int_G q^{\theta g h^{-1}}(y) \mu_G(dg)}$$

$$= \frac{\int_G \eta(\theta g h) q^{\theta g}(y) \Delta_G(gh) \nu_G(dg \cdot h)}{\int_G q^{\theta g}(y) \Delta_G(gh) \nu_G(dg \cdot h)}$$

$$= \frac{\int_G (\eta h)(\theta g) q^{\theta g}(y) \Delta_G(g) \Delta_G(h) \nu_G(dg)}{\int_G q^{\theta g} \Delta_G(g) \Delta_G(h) \nu_G(dg)}$$

$$= \frac{\int_G [(\eta h)(\theta g)] q^{\theta g}(y) \mu_G(dg)}{\int_G q^{\theta g}(y) \mu_G(dg)} = \hat{\eta}_\tau(y) h.$$

Here we have used (3.11), (3.5), (A.21), (A.17) and (A.18).

Proof of Theorem 3.8.2.

It is trivial to verify that the quantities (1)-(5) are invariant under rotation. We have to prove that the set is maximal invariant. To this end we show that if two parameter values $\theta^{(1)}$ and $\theta^{(2)}$ are on the same orbit, then necessarily they have the same values for the quantities (1)-(5).

So assume that $(\Sigma^{(1)}, \beta^{(1)}, \sigma^{(1)})$ and $(\Sigma^{(2)}, \beta^{(2)}, \sigma^{(2)})$ are on the same orbit. Then necessarily $\sigma^{(1)} = \sigma^{(2)}$. Also, there is a rotation C such that $\beta^{(2)} = C\beta^{(1)}$ and $\Sigma^{(2)} = C\Sigma^{(1)}C'$, that is $\Sigma^{(2)}C = C\Sigma^{(1)}$, or

$$\sum_{i=1}^{q^{(2)}} \lambda_i^{(2)} P_i^{(2)} C = \sum_{i=1}^{q^{(1)}} \lambda_i^{(1)} C P_i^{(1)}.$$

This implies

$$\Sigma^{(2)} C P_j^{(1)} = \lambda_j^{(1)} C P_j^{(1)} \text{ and } P_k^{(2)} C \Sigma^{(1)} = \lambda_k^{(2)} P_k^{(2)} C.$$

Multiplying the first equation here from the left by $P_k^{(2)}$ and the second equation from the right by $P_j^{(1)}$ and using again $\Sigma^{(2)} C = C\Sigma^{(1)}$, we find

$$\lambda_j^{(1)} P_k^{(2)} C P_j^{(1)} = \lambda_k^{(2)} P_k^{(2)} C P_j^{(1)}.$$

It follows that either $\lambda_j^{(1)} = \lambda_k^{(2)}$ or $P_k^{(2)} C P_j^{(1)} = 0$. Now fix j. If $P_k^{(2)} C P_j^{(1)} = 0$ for all k, then $I = \sum_{k=1}^{q^{(2)}} P_k^{(2)}$ implies that $C P_j^{(1)} = 0$, hence $P_j^{(1)} = 0$, which is impossible. Therefore there must be at least one k such that $\lambda_k^{(2)} = \lambda_j^{(1)}$. Furthermore, there can not be more than one such k, since the $\lambda_k^{(2)}$ are assumed to be different. Hence if the λ's are ordered according to size, it follows that $\lambda_1^{(1)} = \lambda_1^{(2)}, \ldots, \lambda_q^{(1)} = \lambda_q^{(2)}$ and $q^{(1)} = q^{(2)} = q$.

Finally, $P_k^{(2)} C P_j^{(1)} = 0$ $(j \neq k)$ implies:

$$P_k^{(2)} C \sum_{j \neq k} P_j^{(1)} = P_k^{(2)} C (I - P_k^{(1)}) = 0, \text{ hence } P_k^{(2)} C = P_k^{(2)} C P_k^{(1)}.$$

$$\sum_{k \neq j} P_k^{(2)} C P_j^{(1)} = (I - P_j^{(2)}) C P_j^{(1)} = 0, \text{ hence } C P_j^{(1)} = P_j^{(2)} C P_j^{(1)}.$$

Putting $j = k$ one gets $P_k^{(2)} C = C P_k^{(1)}$, that is $P_k^{(2)} = C P_k^{(1)} C'$ $(k = 1, \ldots, q)$. Since $\beta^{(2)} = C \beta^{(1)}$ it follows in particular that $P_k^{(2)} \beta^{(2)} = C P_k^{(1)} \beta^{(1)}$, and hence $\gamma_k^{(2)} = \gamma_k^{(1)}$ $(k = 1, 2, \ldots, q)$. Also, equality of the dimensions v_k follows.

Proof of Theorem 3.9.1.

(a) It is clear that if $y \mapsto y + c$, where $c \in V = \mathrm{span}(X)$, then $a \mapsto a$, so a is invariant. Since y can be recovered from a and the projection of y upon V, and since no part of this projection can be invariant under translations in V, we must have that a is maximal invariant.

(b) From the model equation of the form $y = X\beta + e$, where e is multinormal $(0, \Sigma)$, we see that $a = Pe$, where $P = I - X(X'X)^{-1}X'$ is the projection upon the $(n - p)$-dimensional space orthogonal to V. From this we see directly that $z = A'e$ has a distribution which is independent of β, specifically it is multinormal $(0, A'\Sigma A)$, which is non-singular, since the covariance matrix here must have rank $n - p$.

(c) For any $n \times (n - p)$ matrix B of rank $n - p$ such that $B'X = 0$ we must have that the columns of B have to span the space orthogonal to V; hence $B = AC$ for a non-singular matrix C. This implies that $B'y = C'A'y$, and the likelihoods of $A'y$ and $B'y$ can be simply transformed into each other.

Proof of Proposition 3.11.2.

Drop the index a. Let T be the space of complete sufficient statistics, and let λ be a measure such that each Q^θ is a measure on T, absolutely continuous with respect to λ. Define for fixed θ and g the set function $R^{\theta,g}(A) = Q^{\theta g}(A)$. Then $R^{\theta,g}$ is a measure.

Define for fixed A the set Ag^{-1} by $R^{\theta,e}(Ag^{-1}) = R^{\theta,g}(A)$. This is uniquely defined by the following argument: Assume that $R^{\theta,e}(B) = R^{\theta,e}(B_1)$ for some set $B_1 = B_1(A) \neq B = B(A)$. Then this implies

$$(I_{B(A)}(du) - I_{B_1(A)}(du)) Q^\theta(du) = 0.$$

Now varying A, this implies two different measures, and since these measures by assumption must be absolutely continuous with respect to each other, it follows that there exists a function k such that

$$\int (f(u) - f(u)k(u))Q^\theta(du) = 0$$

for all functions f. But by (2.58) this implies that $k(u) = 1$ almost surely with respect to the measure $Q^\theta(du)$. In particular, the indicators of the sets B and B' must be equal almost surely with respect to $Q^\theta(du)$.

By a variant of the same argument it follows that $Ag_2^{-1}g_1^{-1} = A(g_1g_2)^{-1}$. Hence the mapping from the actions on Θ to the actions on subsets of T is a homomorphism, and the actions upon T can be seen upon as actions by the same group.

Finally we have

$$Q^\theta(Ag^{-1}) = \int_{Ag^{-1}} R^{\theta,e}(dt)$$

$$= \int_A R^{\theta,g}(dt) = Q^{\theta g}(A)$$

as desired, and uniqueness follows from the results above.

A.2.4 *On Group Representation Theory*

A matrix representation of a group G is defined as a function U from the group to the set of (here complex) matrices satisfying $U(gh) = U(g)U(h)$ for all $g, h \in G$. In other words, a representation is a homomorphism from G to the multiplicative group of square matrices of a fixed dimension. Any representation U and any fixed non-singular matrix K of the same size can be used to construct another representation $S(g) = KU(g)K^{-1}$. If the group is compact (and also in some other cases), we can always find such S of minimal block diagonal form, and at the same time we can take S to be unitary $(S(g)^\dagger S(g) = I)$. If (and only if) the group is Abelian, each minimal block will be one-dimensional.

An important aspect of this reduction appears if we look upon the matrices as operators on a vector space: Then each collection of blocks gives an invariant vector space under the multiplicative group of matrices, and each single minimal block gives an irreducible invariant vector space. For compact groups, the irreducible invariant vector spaces will be finite-dimensional. The minimal matrices in the blocks are called irreducible representations of the group.

More generally, a class of operators $\{U(g); g \in G\}$ (where G is a group) on a, possibly infinite dimensional, vector space is a representation if $U(gh) = U(g)U(h)$ for all g, h. A representation of a compact group has always a complete reduction in minimal matrix representations as described above. In particular, this holds for the unitary regular representation defined on a Hilbert space $L^2(\Phi, \nu)$ by $U_R(g)f(\phi) = f(g^{-1}\phi)$. Here ν is the right invariant measure for G on Φ (defined by $\nu(Bg) = \nu(B)$ for all B, g).

Two useful results are Schur's lemmas:

(1) If U and U' are irreducible representations of different dimensions, and A is such that $U(g)A = AU'(g)$ for all g, then necessarily, $A = 0$.

(2) If U and U' are irreducible representations of the same dimension, and A is such that $U(g)A = AU'(g)$ for all g, then either U and U' are isomorphic or $A = 0$. If $U(g)A = AU(g)$ for all g, then necessarily $A = \lambda I$ for some scalar λ.

An important class of continuous groups, the groups which also are differential manifolds, in such a way that the group operations are compatible with the smooth structure, are the *Lie groups*, which play an important role in modern physics. Simple, but important Lie groups have the representation $U(s) = \exp(isA)$ for some fixed operator A.

More on group representations can be found in Diaconis [62], Hamermesh [102], James and Liebeck [132], Serre [185], Wolbarst [208], Naimark and Štern [159] and Barut and Raczka [19].

A.3   Technical Aspects of Quantum Mechanics

A.3.1   *Parameters of Several Statistical Experiments*

Several places in the main text I have assumed the existence of a c-variable. This subsection gives a very general alternative way to arrive at this concept.

Consider a set $\mathcal{A}$ of mutually exclusive experiments, each of the ordinary statistical kind, but I will concentrate on the parameter spaces $\Lambda^a; a \in \mathcal{A}$. The whole set of parameters of the experiments is given by points in the big space

$$\Pi = \times_a \Lambda^a,$$

a Cartesian product. If all parameter spaces have the same structure Λ, this can be considered to be the set of functions from $\mathcal{A}$ to Λ.

Let there be defined a transformation group G on Π.

Example A1. Let $\pi = (\lambda^1, \lambda^2)$, where λ^1 and λ^2 are the expected lifelengths of a single patient under two mutually exclusive treatments. Let G be the joint set of time scale transformations together with the exchange $\lambda^1 \leftrightarrow \lambda^2$.

Example A2. Consider electron spin. Let $\pi = \times_{a \in \mathcal{A}} \lambda^a$, where λ^a is the spin component of a perfect measurement in the direction a of an electron. Let G be the group generated by the transformations:

i) Inversions: $\lambda^a \mapsto -\lambda^a$.

ii) Rotations of experiments: If $a \mapsto ao$ under a rotation o, replace each λ^a with λ^{ao}, This gives a permutation within the Cartesian product.

Note in general that the points of Π make sense mathematically, but not directly physically. Hence it does not make sense in a concrete physical context to give values to the individual points of this space. And the space will not be called state space.

In particular, I will not look upon it as meaningful to consider Bayesian probabilities over the space Π. Over most subspaces it is also meaningless to talk about Bayesian probabilities, but it is meaningful to consider such probabilities over the λ^a's corresponding to individual experiments.

So what operations are meaningful with the spaces Π? I have mentioned group operations. One can also adjoin such spaces corresponding to different systems, and adjoin π with some other parameter if this is maningful with respect to the theory of the physical system in question. Finally, one can look at subspaces.

Assume that the experiments are related in some way. Then it may be reasonable to try to reduce Π. The purpose of this reduction may be to achieve parsimony. This should not be thought of as an approximation, however, but is a result of some reasonable physical theory. Note that theories are formulated not in terms of observations, but in terms of parameters, the theoretical language behind observations.

Let Π be reduced to a subspace Ψ with the properties:

Property A1. Ψ is an orbit or a set of orbits for the group G. Use the notation G also for this group acting on Ψ.

This is a necessary condition in order that G should be a transformation group on the reduced space. It is also consistent with the discussion elsewhere in this book.

It is natural in certain situations to also demand:

Property A2. Each section $\{\pi \in P : \lambda^a(\pi) = \lambda_0\}$ has a non-zero intersection with Ψ.

In fact, this will always be true for some values λ_0

Let now a model reduction be associated with some function ϕ on Π which is one-to-one on the subset Ψ and undefined elsewhere. It then follows from Property A1 that the group is well defined on the range of ϕ.

Definition A2. *If such a function exists, call* $\Phi = \phi(\Psi)$ *the c-variable space. Any function with the above properties is called a c-variable.*

A c-variable ϕ can in principle be replaced by any other c-variable in one-to-one correspondence with ϕ. But it is important to have a simple representation.

If Property A2 holds, then each λ^a can be regarded as a function on Φ. Furthermore, one can always find a subgroup G^a of G such that this function is permissible under G^a.

Example A2 (continued). Restrict Π to the subset Ψ, the set of all π such that there exists a direction ϕ that gives each λ^a equal to $\text{sign}(a \cdot \phi)$. Let $\phi(\pi)$ be this direction normed as a unit vector.

- The vector $\phi(\pi)$ is a unique function of π. *Proof:* Suppose that there is a π which corresponds to two different vectors ϕ_1 and ϕ_2. Then $a = \phi_1 - \phi_2$, normalized gives $\lambda^a = +1$ corresponding to ϕ_1 and $\lambda^a = -1$ corresponding to ϕ_2, a contradiction.

- The set Ψ is an orbit of G. *Proof:* It is easy to see that Ψ is closed under inversions and rotations.

- All sections $\{\pi : \lambda^a = \pm 1\}$ have nonzero intersections with Ψ. *Proof:* Obvious.

- The subgroup G^a as defined on Φ is generated by the set of rotations around a together with a single rotation of $180°$ around an axis perpendicular to a. Proved before.

A.3.2   *Proofs from Section 5.4*

Proof of Theorem 5.4.2.

Without loss of generality consider a system where each experimental parameter λ only takes two values, say -1 and $+1$. Otherwise we can reduce to a degenerate case with just these two eigenvalues: The statement $v_i^a = v_j^b$ only concerns the eigenvalues λ_i^a and λ_j^b; without loss of generality the other eigenvalues can be taken as equal.

Take $U^a = W(g_{ca})U(g_{ac})$ and $U^b = W(g_{cb})U(g_{bc})$. The assumption is that

$$U^a f_i^a = U^b f_j^b. \tag{A.23}$$

Using (5.8) and its extensions, we see that every $W(g)$ can be written as a product of $U()$–terms, and therefore as some $U(g')$. This implies that U^a is equal to some $U(g'')$ and similarly U^b, so finally $(U^b)^{(-1)}U^a$ is equal to some $U(g_0)$ for the right regular representation $U()$.

Without loss of generality, let both λ_i^a and λ_j^b be equal to 1. Then

$$I(\lambda^b(\phi) = 1) = f_j^b(\phi) = (U^b)^{(-1)}U^a f_i^a(\phi) = I((\lambda^a g_0)(\phi) = 1).$$

Thus $I(\lambda^b = 1) = I(\lambda^a g_0 = 1)$. But this means that the two level curves coincide, and we must have $\lambda^b = \lambda^a g_0 = F(\lambda^a)$.

(b) follows trivially from (a).

Proof of Theorem 5.10.1.

We will show that the mapping A^a can be replaced by a unitary map in the relation $\mathcal{H}^a = A^a \mathcal{K}^a$.

Recall from Definition 3.2.1 that the connection $\dot{g} \mapsto \bar{g}$ from the observation group to the parameter group $\bar{G}^a$ is given from the reduced model by

$$P^{\lambda^a \bar{g}}(B) = P^{\lambda^a}(B\dot{g}^{-1}). \tag{A.24}$$

For $\dot{g} \in \dot{G}$ and $\bar{g} \in \bar{G}^a$ define $U_1(\dot{g}) = \bar{U}(\bar{g})$ as operators on $\mathcal{H}^a$ when $\dot{g} \mapsto \bar{g}$ as in (A.24). Here $\bar{U}$ is the regular representation of the group $\bar{G}^a$. Then it is easy to verify that U_1 is a representation of $\dot{G}$. Also, if V_1 is an invariant space for U_1, then it is also an invariant space for $\bar{U}$. However, the space V_1 is not necessarily irreducible for $\bar{U}$ even if it is irreducible for U_1.

Using the definition (5.33) and the connection (A.24) between $\dot{g}$ and $\bar{g}$ we find the following relationships. We assume that the random variable $y(\cdot)$ belongs to $\mathcal{K}^a \subset L^2(S, \mathrm{P})$ and that $\bar{U}$ is chosen as a representation on the invariant space $\mathcal{H}^a$. Then

$$
\begin{aligned}
U_1(\dot{g})A^a y(\lambda^a) &= \bar{U}(\bar{g})A^a y(\lambda^a) = \int y(\omega)\mathrm{P}^{\lambda^a \bar{g}}(d\omega) \\
&= \int y(\omega)\mathrm{P}^{\lambda^a}(d\omega \dot{g}^{-1}) = \int y(\omega \dot{g})\mathrm{P}^{\lambda^a}(d\omega) = A^a \dot{U}(\dot{g})y(\lambda^a),
\end{aligned}
\tag{A.25}
$$

where $\dot{U}$ is the representation on $\mathcal{K}^a$ given by $\dot{U}y(\omega) = y(\omega \dot{g})$, i.e., the regular representation on $L^2(S, \mathrm{P})$ restricted to this space.

Thus $U_1(\dot{g})A^a = A^a \dot{U}(\dot{g})$ on $\mathcal{K}^a$.

Furthermore

$$
U(g) = \bar{U}(\bar{g}) = U_1(\dot{g}) = A^a \dot{U}(\dot{g})A^{a\,-1} \text{ when } \dot{g} \mapsto \bar{g} \text{ and } g \mapsto \bar{g}.
$$

Recall that $g \mapsto \bar{g}$ in this setting if $(\lambda^a \bar{g})(\phi) = \lambda^a(\phi g)$, and that $\bar{U}(\bar{g}) = U(g)$ in this case. Furthermore, $U(g)f(\phi) = f(\phi g)$ when $f \in V_\lambda^a$ and $g \in G^a$.

By [159] p. 48, if two representations of a group are equivalent, they are unitary equivalent; hence for some unitary C^a we have

$$
\bar{U}(\bar{g}) = C^a \dot{U}(\dot{g})C^{a\,\dagger}
\tag{A.26}
$$

when $\dot{g} \mapsto \bar{g}$.

Since the unitary operators in this proof are defined on $\mathcal{K}^a$ and $\mathcal{H}^a$, respectively, it follows that these spaces are related by $\mathcal{H}^a = C^a \mathcal{K}^a$.

Definition 5.10.2 may also be coupled to the operator A^a and to an arbitrary Hilbert space $\mathcal{K}'$ of sufficient statistics, which may trivially be the whole space $L^2(S, \mathrm{P})$. Let first

$$
L^a = \{y \in \mathcal{K}' : \mathrm{E}^{\lambda^a} y = 0 \text{ for all } \lambda^a\}.
\tag{A.27}
$$

Then $\mathcal{K}^a$ may be considered as the factor space $\mathcal{K}'/L^a$, i.e., the equivalence classes of the old $\mathcal{K}'$ with respect to the linear subspace L^a (cf [159], I.2.10V).

Here is a proof of this fact: Let $\xi \in A^a \mathcal{K}'$, such that $\xi(\lambda^a) = \mathrm{E}^{\lambda^a}(y)$ for some $y \in \mathcal{K}'$. Then y is an unbiased estimator of the function $\xi(\lambda^a)$. By [146], Lemma 1.10, $\xi(\lambda^a)$ has one and only one unbiased estimator which is a function $h(t^a)$ of t^a. Then every unbiased estimator of $\xi(\lambda^a)$ is of the form $y = h(t^a) + x$, where $x \in L^a$; this constitutes an equivalence class. On the other hand, every $h(t^a)$ can be taken as such a y.

A.4   Some Aspects of Partial Least Squares Regression

Lemma A4.1 *The weights $w_k = E'_{k-1}f_{k-1}$ lead to a maximisation of the empirical covariance between t_k and f_{k-1}.*

<u>Proof.</u>

Since all variables are centered, the empirical covariance is

$$f'_{k-1} t_k = f'_{k-1} E_{k-1} w_k,$$

and for a fixed norm of the weight w_k this is maximized for $w_k = E'_{k-1} f_{k-1}$.

An alternative PLS algorithm. *As a start, put $E_0^* = X$, $f_0^* = y$, and then determine $p_k^*, t_k^*, T_k^*, \tilde{q}_k^* = (q_{k1}^*, ..., q_{kk}^*)', E_k^{*'}$ and f_k^* consecutively by the formulas*

$$p_k^* = E_{k-1}^{*'} f_{k-1}^*,$$

$$t_k^* = E_{k-1}^* p_k^* / p_k^{*'} p_k^*,$$

$$T_k^* = (t_1^*, ..., t_k^*),$$

$$\tilde{q}_k^* = (T_k^{*'} T_k^*)^{-1} T_k^{*'} y,$$

$$E_k^* = E_{k-1}^* - t_k^* p_k^{*'},$$

$$f_k^* = y - \sum_{j=1}^{k} t_j^* q_{kj}^*.$$

If again $x_0 = (x_{01}, ..., x_{0p})'$ is a set of x-measurements on a new unit, define $e_0^* = x_0 - \bar{x}$ with $\bar{x} = (\bar{x}_1, ..., \bar{x}_p)'$ and then new scores and residuals consecutively by

$$t_{k0}^* = e_{k-1}^{*'} p_k^* / p_k^{*'} p_k^*,$$

$$e_k^* = e_{k-1}^* - t_{k0}^* p_k^* = e_0^* - \sum_{j=1}^{k} t_{j0}^* p_j^*.$$

Predict y_0 in step a by

$$\hat{y}_{a0}^* = \bar{y} + \sum_{k=1}^{a} t_{k0}^* q_{ak}^*.$$

The next result shows to which extent the above algorithm is equivalent to the algorithm in Section 8.1:

Proposition A4.1. *With the notation as above we have for $a = 1, 2, ...$:*
a) $p_a^* = w_a$,
b) $\{t_1^*, ..., t_a^*\}$ span the same space as $\{t_1, ..., t_a\}$,
c) $f_a^* = f_a$,
d) $\hat{y}_{a0}^* = \hat{y}_{a0}$.

Proof.

Let P_{ta}^* and P_{ta} be the projections upon the two spaces described in b). Then using the orthogonality of $t_1, t_2, ...,$ we find from $(8.2) - (8.5)$:

$$E_a = (I - P_{ta})X,$$

$$f_a = (I - P_{ta})y,$$

while the above algorithm gives

$$E_a^* = X - \sum_{k=1}^{a} t_k^* p_k^{*'},$$

$$f_a^* = (I - P_{ta}^*)y.$$

I will prove a) and b) simultaneously by induction on a. They are trivial for $a = 1$. Assume them to be true up to $a - 1$. Then the formulas above show that

$$E_{a-1}' f_{a-1} = E_{a-1}^{*'} f_{a-1}^* = X'(I - P_{t,a-1})y,$$

giving a). But the two set of scores are given by

$$t_a = E_{a-1}w_a = (I - P_{t,a-1})Xw_a = Xw_a - P_{t,a-1}Xw_a,$$

$$t_a^* \propto E_{a-1}^* p_a^* = (X - \sum_{k=1}^{a-1} t_k^* p_k^{*'})p_a^* = Xp_a^* = Xw_a,$$

where the last term vanishes because of the othogonality. Since, by the induction hypothesis, the vector subtracted from Xw_a in t_a belongs to the span of $t_1^*, ..., t_{a-1}^*$, this proves b).

By the formulas above for f_a and f_a^*, c) follows from b).

It follows from b) that $T_a^* = T_a D$ for some non-singular matrix D, where $T_a = \{t_1, ..., t_a\}$. From the way the scores t_{k0}^* and t_{k0} are constructed, one must then also have $[t_{10}^*, ..., t_{a0}^*] = [t_{10}, ..., t_{a0}]D$. Thus we have a simple linear change of variable between the two sets of scores, and since prediction in both cases is based upon linear regression on the scores, d) follows.

Proposition A4.2. *We have the following relationships:*

a) $w_{a+1} = s - SW_a(W_a'SW_a)^{-1}W_a's$, *where* $s = X'y, S = X'X$ *and* $W_a = (w_1, ..., w_a)$.

b) $\hat{y}_{a0} = \bar{y} + (x_a - \bar{x})'b_a$ *with* $b_a = W_a(W_a'SW_a)^{-1}W_a's$.

Proof.

From the previous proof, t_a^* is proportional to Xw_a, or properly normed $t_a^* = Xw_a/w_a'w_a$. Thus $T_a^* = XW_aC_a$, where $C_a = \text{diag}(\|w_1\|^{-2}, ..., \|w_a\|^{-2})$. From this

$$w_{a+1} = E_a^{*'} f_a^* = X'(I - P_{ta}^*)y$$

with $P_{ta}^* = T_a^*(T_a^{*'}T_a^*)^{-1}T_a^{*'}$ gives a).

From d) in Proposition A4.1 and from the definition of $\hat{y}_{a0}^*$ we find

$$\hat{y}_{a0} = \bar{y} + (t_{10}^*, ..., t_{a0}^*)\tilde{q}_a^*.$$

Similarly to the relation $T_a^* = XW_aC_a$, we get

$$(t_{10}^*, ..., t_{a0}^*) = (x_0 - \bar{x})'W_aC_a.$$

Finally, the definition of $\tilde{q}_a^*$ gives

$$\tilde{q}_a^* = C_a^{-1}(W_a'SW_a)^{-1}W_a's.$$

b) follows.

Bibliography

[1] Accardi, L.: The quantum probabilistic approach to the foundation of quantum theory. In: Urns and chameleons. Dalla Chiara. M.L. et al. (eds.) Language, Quantum, Music, 95-104. Kluwer Academic Publishers (1999)

[2] Accardi, L., Regoli, M.: The EPR correlation and the chameleon effect. http://xxx.lanl.gov/abs/quant-ph/0110086 (2001)

[3] Accardi, L., Imafuku, K., Regoli, M.: On the physical meaning of the EPR-chameleon experiment. http://xxx.lanl.gov/abs/quant-ph/0112067 (2001)

[4] Accardi, L.: Non relativistic quantum mechanics as a noncommutative Markov process. Adv. in Math. **20**, 329-366 (1976)

[5] Accardi, L. and Heyde, C.C.: Probability Towards 2000. Springer-Verlag, New York (1998)

[6] Aerts, D., S. Aerts, Broekaert, J. and Gabora, L.: The violation of Bell inequalities in the macroworld. Found. Phys. **30**, 1387-1414 (2000)

[7] Aerts, D., Czachor, M. and D'Hooghe, B.: Towards a quantum evolutionary scheme: violating Bell's inequalities in language. In: Gontier, N, Van Bendegem, J.P. and Aerts, D. (Eds.) Evolutionary Epistemology, Langage and Culture. Studies in Language, Companion series. John Benjamins Publishing Company (2005)

[8] Aerts, D. and Gabora, L.: A theory of concepts and their combinations I. Kybernetes **34**, 167-191 (2005)

[9] Aerts, D. and Gabora, L.: A theory of concepts and their combinations II. Kybernetes **34**, 192-221 (2005)

[10] Aitkin, M.: Comments to Helland, I.S. (1995) Simple counterexamples against the conditionality principle. Amer. Statist. **50**, 384-385 (1996)

[11] Almøy, T.: A simulation study on comparison of prediction methods when onl a few components are relevant. Comput. Stat. Data Anal. **21**, 87-107 (1996)

[12] Arnold, S.F.: Sufficient statistics. In: Kotz, S., Johnson, N.L. and Read, C.B. Encyclopedia of Statistical Sciences. Wiley, New York (1988)

[13] Bailey, R.A.: A unified approach to design of experiments. J. R. Statist. Soc. A **144**, 214-223 (1981)

[14] Bailey, R.A.: Strata for randomized experiments. J. R. Statist. Soc. B **53**, 27-78 (1991)

[15] Bailey, R.A.: Design of Comparative Experiments. Cambridge University Press (to appear)

[16] Barnum, H., Caves, C.M., Finkelstein, J., Fuchs, C.A. and Schack, R.: Quantum probability from decision theory? Proc. R. Soc. Lond. **A 456**, 1175-1182 (2000)

[17] Barndorff-Nielsen, O.E. and Gill, R.D.: Fisher Information in Quantum Statistics. J. Phys. A: Math. Gen. **33**, 4481-4490 (2000)

[18] Barndorff-Nielsen, O.E., Gill, R.D. and Jupp, P.E.: On quantum statistical inference (with discussion). J. R. Statist. Soc. B **65**, 775-816 (2003)

[19] Barut, A.S., Raczka R.: Theory of Group Representation and Applications. Polish Scientific Publishers, Warsaw (1985)

[20] Bateson, G.: Steps to an Ecology of Mind. University of Chicago Press, Chicago (2000)

[21] Belavkin, V.P.: Quantum probabilities and paradoxes of the quantum century. Infinite Dimensional Analysis, Quantum Probability and Related Topics **3**, 577-610 (2000)

[22] Bell, J.S.: On the problem of hidden variables in quantum mechanics. Rev. Mod. Physics **38**, 447-452 (1966)

[23] Bell, J.S.: Berlmann's socks and the nature of reality. In: Speakable and unspeakable in quantum mechanics. Cambridge University Press, Cambridge (1987)

[24] Berger, J.O.: Statistical Decision Theory and Bayesian Analysis. Springer-Verlag, New York (1985)

[25] Berger, J.O., and Wolpert, R.L.: The Likelihood Principle. Lecture Notes, Monograph Series, Vol. 6, Institute of Mathematical Statistics, Hayward (1984)

[26] Bernardo, J.M. and Smith, A.F.M.: Bayesian Theory. Wiley, Chichester (1994)

[27] Bickel, P.J., Doksum, K.A.: Mathematical Statistics. Basic Ideas and Selected Topics. Vol. I. Prentice-Hall, Upper Saddle River, NJ (2001)

[28] Birnbaum, A.: On the foundation of statistical inference. J. Amer. Statist. Ass. **57**, 269-326 (1962)

[29] Bohm, D.: A suggested interpretation of the quantum theory in terms of "hidden" variables. Int. Phys. Rev. **85**, 166-179 (1952)

[30] Bohr, A. and Ulfbeck, O.: Primary manifistation of symmetry. Origin of quantal indeterminacy. Rev. Mod. Phys. **67**, 1-35 (1995)

[31] Bondar, J.V.: Structural distributions without exact transitivity. Ann. Math. Statist. **43**, 326-339 (1972)

[32] Bondar, J.V., Milnes, P.: Amenability: A survey for statistical applications of Hunt-Stein and related conditions on groups. Z. Wahrscheinlichkeitstheorie verw. Gebiete **57**, 103-128 (1981)

[33] Bondesson, L.: Equivariant estimators. In: Kotz, S., N.L. Johnson and C.B. Read (ed.) Encyclopedia of Statistical Sciences. Wiley, New York (1982)

[34] Box, G.E.P., Hunter, J.S. and Hunter, W.G.: Statistics for Experimenters. Wiley, New York (2005)

[35] Braunstein, S.L. and Caves, C.M.: Statistical distance and the geometry of quantum states. Phys. Review Letters **72**, 3439-3443 (1994)

[36] Breiman, L.: Statistical modelling: The two cultures. Statistical Science **16**, 199-231 (2001)

[37] Burnham, K.P., Anderson, D.R.: Model Selection and Multimodel Inference. A Practical Information-Theoretic Approach. Springer-Verlag, New York (2002)

[38] Busch, P.: Quantum states and generalized observables: A simple proof of Gleason's theorem. Phys. Rev. Letters 91 (12) (2003)

[39] Busch, P., Heinonen, T. and Lahti, P.: Heisenberg's uncertainty principle. In: 15th UK and European Meeting on the Foundations of Physics, Leeds (2007)

[40] Cartwright, N.: How the Laws of Physics Lie. Clarendon Press, Oxford (1983)

[41] Caves, C.M., Fuchs, C.A., Manne, K. and Renes, J.M.: Gleason-type derivations of the quantum probability rule for generalized measurements. arXiv:quant-ph/0306179 (2003)

[42] Cheng, B. and Wu, X.: A modified PLSR method in prediction. J. Data Science **4**, 257-274 (2006)

[43] Clarke, P.: Towards a greater understanding of the experience of stroke: Integrating qualitative and quantitative methods. J. Aging Studies **17**, 171-187 (2003)

[44] Comte, A.:A General View Of Positivism (1856)

[45] Cook, R.D., Li, B. and Chiaromonte, F.: Dimension reduction in regression without matrix inversion. Preprint. (2007)

[46] Cook, R.D., Li, B. and Chiaromonte, F,: Envelope models for parsimonious and efficient multivariate linear regression. Statistica Sinica. To appear with discussion (2009)

[47] Corbeil, R.R, Searle, S.R.: Restricted maximum likelihood (REML) estimation of variance components in the mixed model. Technometrics **18**, 31-38 (1976)

[48] Cox, D.R.: Some problems connected with statistical inference. Ann. Statist. **29**, 357-372 (1958)

[49] Cox, D.R.: The choice between ancillary statistics. J. Roy. Statist. Soc. B **33**, 251-255 (1971)

[50] Cox, D.R. and Hinkley, D.V.: Theoretical Statistics. Chapman and Hall, London (1974)

[51] Dawid, A.P.: Invariant prior distributions. In: Kotz, S., Johnson, N.L., Read, C.B.: Encyclopedia of Statistical Sciences. John Wiley, New York (1983)

[52] Dawid, A.P.: Symmetry models and hypotheses for structural data layouts. J. R. Statist. Soc. B **50**, 1-34; Ann. Statist. **10**, 1054-1067 (1988)

[53] Dawid, A.P.: Causal inference without couterfactuals. JASA **95**, 407-448 (2000)

[54] Dawid, A.P., Stone, M., Zidek, J.V.: Marginalization paradoxes in Bayesian and structural inference, J. R. Statist. Soc. B **35**, 189-233 (1973)

[55] Dawid, A.P., Stone, M., Zidek, J.V.: Critique of E.T. Jaynes's "Paradoxes of Probability Theory". Unpublished manuscript.

[56] de Jong, S.: PLS fits closer than PCR. J. Chemom. **7**, 551-557 (1993)

[57] de Jong, S.: SIMPLS: an alternative approach to partial least squares regression. Chemometrics and Intelligent Laboratory Systems **18**, 251-263 (1993)

[58] de Jong, S.: PLS shrinks. J. Chemom. **9**, 323-326 (1995)

[59] Deutsch, D.: The Fabric of Reality. Allen Lane. The Penguin Press (1997)

[60] Deutsch, D.: Quantum theory of probability and decisions. Proc. Roy. Soc. **A 455**, 3129-3197 (1999)

[61] De Witt, B.S.: The many-universes interpretation of quantum mechanics. In: Foundations of Quantum Mechanics. IL Corso. Academic Press, New York (1971)

[62] Diaconis, P.: Group Representations in Probability and Statistics. IMS Lecture Notes. Monograph Series, Vol. 11. Hayward, California (1988)

[63] Dirac, P.A.M.: The Principles of Quantum Mechanics. The Clarendon Press, Oxford (1930)

[64] Döring, A. and Isham, C.J.: A topos foundation for theories of of physics: I. Formal language for physics. arXiv:quant-ph/070360 (2007)

[65] Döring, A. and Isham, C.J.: A topos foundation for theories of of physics: II. Daseinisation and the liberation of quantum theory. arXiv:quant-ph/070362 (2007)

[66] Döring, A. and Isham, C.J.: A topos foundation for theories of physics: III. The representation of physical quantities with arrows. arXiv:quant-ph/070364 (2007)

[67] Döring, A. and Isham, C.J.: A topos foundation for theories of of physics: IV. Categories of systems. arXiv:quant-ph/070366 (2007)

[68] Eaton, M.L.: Group Invariance Applications in Statistics. Institute of Mathematical Statistics and American Statistical Association, Hayward, California (1989)

[69] Eaton, M.L., Sudderth, W.D.: Prediction in a multivariate setting: Coherence and incoherence. Sankhya A **55**, 481-493 (1993)

[70] Eaton, M.L., Sudderth, W.D.: The formal posterior of a standard flat prior in MANOVA is incoherent. J. Italian Statist. Soc. **2**, 251-270 (1995)

[71] Eaton, M.L., Sudderth, W.D.: A new predictive distribution for normal multivariate linear models. Sankhya A **60**, 363-382 (1998)

[72] Eaton, M.L., Sudderth, W.D.: Consistency and strong inconsistency of group-invariant predictive inferencees. Bernoulli **5**, 833-854 (1999)

[73] Eaton, M.L., Sudderth, W.D.: Group invariant inference and right Haar measures. J. Stat. Planning and Infer. **103**, 87-99 (2002)

[74] Einstein, A., Podolski, B., Rosen, N.: Can quantum-mechanical description of physical reality be considered complete? Phys. Rev. **47**, 777-780 (1935)

[75] Efron, B.: R.A. Fisher in the 21st century. Statistical Science **13**, 95-122 (1998)

[76] Evans, M., Fraser, D.A.S., Monette, G: On principles and arguments to likelihood. Canad. J. Statist. **14**, 181-199 (1986)

[77] Everett, H. III: "Relative state" formulation of quantum mechanics, Rev. Modern Physics **29**, 454-462 (1957)

[78] Everett, H. III: The theory of the universal wave function. In: The Many-Worlds Interpretation of Quantum Mechanics. DeWitt, B.S. and Graham, N. (Ed.) Princeton University Press, Princeton (1973)

[79] Feller, W.: An Introduction to Prrobability Theory and its Applications. Vol. 1. Wiley, New York (1950)

[80] Feller, W.: An Introduction to Probability Theory and its Applications. Vol. 2. Wiley, New York (1966)

[81] Finkelstein, J.: Quantum probability from decision theory? http://arxiv.org/abs/quant-ph/9907004 (1999).

[82] Fisher, R.A.: On the mathematical foundation of theoretical statistics. Reprinted together with other papers in R.A. Fisher (1950) Contribution to Mathematical Statistics. Wiley, New York (1922)

[83] Fisher, R.A.: Statistical Methods fro Research Workers. Oliver and Boyd, Edinburgh (1925)

[84] Fisher, R.A.: The Design of Experiments. Oliver and Boyd, London (1935)

[85] Frank, I.E., Friedman, J.H.: A statistical view of some chemometrics regression tools (with discussion). Technometrics **35**, 109-148 (1993)

[86] Fraser, D.A.S.: The fiducial method and invariance. Biometrika **48**, 261-280 (1961)

[87] Fraser, D.A.S.: The Structure of Inference. Wiley, New York (1968)

[88] Fraser, D.A.S.: Inference and Linear models. McGraw-Hill, New York (1979)

[89] Freund, J.E.: Mathematical Statistics with Applications. Pearson Education, Upper Saddle River, NJ (2004)

[90] Frieden, B.R.: Physics from Fisher Information. Cambridge University Press, Cambridge (1998)

[91] Gamerman, D.: Markov Chain Monte Carlo: Stochastic Simulation for Bayesian Inference. Chapman & Hall (1997)

[92] Gell-Mann, M. and Hartle, J.B.: Alternatively decohering histories in quantum mechsnics. In: Complexity, Entropy, and the Physics of Information. Ed.: W.H. Zurek. Addison-Wesley (1991)

[93] Gill, R.D., Time, finite statistics, and Bell's fifth position. In: Foundations of Probability and Physics - 2, Vol. 5 of Math. Modelling, Phys., Engin. and Cogn. Sc., Växjö University Press (2003)

[94] Gill, R.: Accardi contra Bell (cum mundi): The impossible coupling. In: M. Moore,

S. Froda and C. Láger (eds.) Mathematical Statistics and Applications. Festschrift for Constance von Eeden; IMS Lecture Notes - Monograph **42**, 133-154. Institute of Mathematical Statistics, Hayward, Ca. (2002)

[95] Gill, R.D.: On an argument of David Deutsch. In: M. Schürmann and U. Franz [Ed.] QP-PQ. Quantum Probability and White Noise Analysis - Vol. 18. Quantum Probability and Infinite Dimensional Analysis. From Foundation to Applications. World Scientific (2005)

[96] Gill, R.D. and Robins, J.M.: Causal inference for complex longitudinal data. Ann. Statist. **29**, 1785-1811 (2001)

[97] Gleason, A.: Measures on closed subspaces of a Hilbert space. Journal of Mathematics and Mechanics **6**, 885-893 (1957)

[98] Goutis, C.: Partial least squares algorithm yields shrinkage estimators. Ann. Statist. **24**, 816-824 (1996)

[99] Goyal, P.: An information-theoretic approach to quantum theory, I: The abstract quantum formalism. http://arxiv.org/abs/quant-ph/0702124v1 (2007).

[100] Griffiths, R.B.: Consistent histories and the interpretation of quantum mechanics. J. Statistical Physics **36**, 219-272 (1984)

[101] Gödel, K.: On Formally Undecidable Propositions of Principia Mathematica and Related Systems. Dover (1962)

[102] Hamermesh, M.: Group Theory and its Application to Physical Problems. Addison-Wesley, Reading, Massachusetts (1962)

[103] Hastie, T., Tibshirani, R. and Friedman, J.: The Elements of Statistical Learning. Springer, New York (2001)

[104] Helland, I.S.: On the structure of partial least squares regression. Commun. Stat. Simul. **17**, 581-607 (1988)

[105] Helland, I.S.: Partial least squares regression and statistical models. Scand. J. Statist. **17**, 97-114 (1990)

[106] Helland, I.S.: Maximum likelihood regression on relevant components. J. R. Statist. Soc. B **54**, 637-647 (1992)

[107] Helland, I.S.: Simple counterexamples against the conditionality principle. Am. Statistician **49**, 351-356 (1995); discussion: **50**, 382-386 (1996)

[108] Helland, I.S.: Restricted maximum likelihood from symmetry. Statistical Research Report No. 14/1999. Department of Mathematics, University of Oslo. Available at http://folk.uio.no/ingeh/publ.html (1999)

[109] Helland, I.S.: Reduction of regression models under symmetry. Invited contribution to: Viana, M. and D. Richards (ed.) Algebraic Methods in Statistics. Contemporary Mathematics Series of the American Mathematical Society (2001)

[110] Helland, I.S.: Some theoretical aspects of partial least squares regression. Chemometrics and Intelligent Laboratory Systems **58**, 97-107 (2001)

[111] Helland, I.S.: Statistical inference under a fixed symmetry group. Preprint (2002)

[112] Helland, I.S.: Discussion of McCullagh, P. (2002) What is a statistical model? Ann. Statistics **30**, 1225-1310 (2002)

[113] Helland, I.S.: Extended statistical modelling under symmetry: The link towards quantum mechanics. Ann. Statist. **34**, 42-77 (2006)

[114] Helland, I.S.: Quantum theory as a statistical theory under symmetry and complementarity. Preprint (2003)

[115] Helland, I.S.: Statistical inference under symmetry. Int. Statist. Rev. **72**, 409-422 (2004)

[116] Helland; I.S.: Quantum mechanics from focusing and symmetry. Found. Phys. **38**, 818-842 (2008)

[117] Helland, I.S., Almøy, T. Comparison of prediction methods when only a few components are relevant. J. Am. Stat. Assoc. **89**, 583-591 (1994)

[118] Helland, I.S. and Cook, R.D.: Partial least squares regression from envelope models. In preparation (2009)

[119] Helstrom, C.W.: Minimum mean-square error of estimates in quantum statistics. Phys. Letters **25** A, 101-102 (1967)

[120] Helstrom, C.W.: Quantum Detection and Estimation Theory. Academic Press, New York (1976)

[121] Hewitt, E., Ross, K.A.: Abstract Harmonic Analysis, II. Springer-Verlag, Berlin (1970)

[122] Hjort, N.L.: Estimation in moderately misspecified models. Statistical Research Report No. 8, Department of Mathematics, University of Oslo (1990)

[123] Hjort, N.L., Claeskens, G.: Frequentist Model Average Estimators. J. Amer. Statist. Ass. **98**, 879-899 (2003)

[124] Holcomb, T.R., Hjalmarsom, H., Morani, M. and Tyler, M.L.: Significant regression: a statistical approach to partial least squares. J. Chemometrics **11**, 283-309 (1997)

[125] Holevo, A.S.: Probabilistic and Statistical Aspects of Quantum Theory. North-Holland Publishing Company, Amsterdam (1982)

[126] Holevo, A.S.: Statistical Structure of Quantum Theory. Springer, Berlin (2001)

[127] Hora, R.B., Buehler, R.J.: Fiducial theory and invariant estimation. Ann. Math. Statist. **37**, 643-656 (1966)

[128] Høglend, P., Amlo, S., Marble, A., Bøgwald, K.-P., Sørby, Ø, Sjaastad, M.C., Heyerdahl, O.: Analysis of the patient-therapist relationship in dynamic psychotherapy: An experimental study of transference interpretation. Am. J. Psychiatry **163** (2006)

[129] Höskuldson, A.: PLS regression methods. J. Chemometrics, **2**, 211-220 (1988)

[130] Isham, C.J.: Lectures on Quantum Theory. Imperial College Press, London (1995)

[131] Jaynes, E.T.: Probability Theory. The Logic of Science. Cambridge University Press, Cambridge (2003)

[132] James, G. and Liebeck, M.: Representation and Characters of Groups. Cambridge University Press, Cambridge (1993)

[133] Kariya, T.: Equivariant estimation in a model with an ancillary statistic. Ann. Statist. **17**, 920-928 (1989)

[134] Kass, R.E., Wasserman, L.: The selection of prior distributions by formal rules. J. Amer. Stat. Ass. **91**, 1343-1370 (1996)

[135] Khrennikov, A.: Contextual viewpoint to quantum stochastics. http://xxx.lanl.gov/abs/hep-th/0112076 (2001)

[136] Khrennikov, A.: Ensemble fluctuation and the origin of quantum probabilistic rule. J. Math. Phys. **43**, 789-802 (2002)

[137] Kolmogorov, A.N.: Grundbegriffe der Wahrscheinlichkeitsrechnung. Springer-Verlag. Berlin (1933)

[138] Kuhn, T.S.: The Structure of Scientific Revolutions (2nd ed.) University of Chicago Press, Chicago (1970)

[139] Kullback, S.: Information Theory and Statistics. Wiley, New York (1959)

[140] Laloë, F.: Do we really understand quantum mechanics? Strange correlations, paradoxes and theorems. http://arxiv.org/abs/quant-ph/0209123 (2002)

[141] Lancaster, H.O.: Statistics, history of. In: Kotz, S., Johnson, N.L., Read, C.B.: Encyclopedia of Statistical Sciences. Wiley, New York (1988)

[142] Larsson, J.-Å. and Gill, R.D.: Bell's inequality and the coincidence-time loophole. Europhysics Letters **67**, 707-713 (2004)

[143] Lauritzen, S.: Graphical Models. Clarendon, Oxford (1996)

[144] Lehmann, E.L.: Testing Statistical Hypotheses. Wiley, New York (1959)

[145] Lehmann, E.L.: Statistics, an overview. In: Kotz, S., Johnson, N.L., Read,C.B.: Encyclopedia of Statistical Sciences. Wiley, New York (1988)

[146] Lehmann, E.L., Casella, G.: Theory of Point Estimation. Springer, New York (1998)

[147] Lehmann, E.L., Scheffé,H.: Completeness, similar regions, and unbiased estimation. Sankhya **10**, 305-340; **15**, 219-236; correction **17**, 250 (1950, 1955, 1956)

[148] Lockwood, M.: 'Many minds'. Interpretations of quantum mechanics. British J. Philos. Sci. **47**, 159-188 (1996)

[149] Malley, J.D. and Hornstein, J.: Quantum Statistical Inference. Statistical Science **8**, 433-457 (1993)

[150] Martens, H.: Multivariate Calibration. Dr. Techn.Thesis, Technical University of Norway, Trondheim, Norway (1985)

[151] Martens, H. and Næs, T.: Multivariate Calibration. Wiley, New York (1989)

[152] McCullagh, P. and Nelder, J.: Generalized Linear Models. Chapman and Hall, London (1989)

[153] Miller, R.G.: Simultaneous Statistical Inference. Springer-Verlag, New York (1981)

[154] Montgomery, D.C.: Design and Analysis of Experiments. Wiley, New York (1997)

[155] McCullagh, P.: Linear models, vector spaces, and residual likelihood. In: Gregoire et al. (eds.): Modelling Longitudinal and Spatially Correlated Data. Springer Lecture Notes No. 122, pp. 1-10 (1996)

[156] McCullagh, P.: What is a statistical model? Ann. Statistics **30**, 1225-1310 (2002)

[157] Meyer, P.-A.: Quantum Probability for Probabilists. Springer-Verlag, Berlin (1995)

[158] Nachbin, L.: The Haar Integral. Van Nostrand, Princeton, N.J. (1965)

[159] Naimark, M.A., Štern, A.I.: Theory of Group Representations. Springer-Verlag (1982)

[160] Nelder, J.A.: A re-formulation of linear models (with discussion). J. Roy. Statist. Soc. Ser. A **140**, 48-77 (1977)

[161] Nelder, J.A.: The combination of information in generally balanced designs. J. Roy. Stat. Soc. Ser. B **30**, 303-311 (1968)

[162] von Neumann, J.: Matematische Grundlagen der Quantenmechanik. Springer, Berlin (1932)

[163] von Neumann, J.: Mathematical Foundations of Quantum Mechanics. Princeton University Press, Princeton (1955)

[164] Neyman, J.: On the application of probability theory to agricultural experiments. Essay on Principles. Section 9. Translated in: Statistical Science (1990) **5**, 465-480 (1923)

[165] Neyman, J., Scott, E.L.: Consistent estimators based on partially consistent observations. Econometrica **16**, 1-32 (1948)

[166] Næs, T. and Helland, I.S.: Relevant components in regression. Scand. J. Statistics **20**, 239-250 (1993)

[167] Omnès, R.: Understanding Quantum Mechanics. Princeton University Press, Princeton (1999)

[168] O'Neill, P.: Tectonic Change: The qualitative paradigm in psychology. Canad. Psychol. **43**, 190-194 (2001)

[169] Parthasarathy, K.R.: An Introduction to Quantum Stochastic Calculus. Birkhuser Verlag, Basel (1992)

[170] Patterson, D., Thompson, R.: Recovery of inter-block information when block sizes are unequal. Biometrika **58**, 545-554 (1971)

[171] Pearl, J.: Causality. Cambridge University Press (2000)

[172] Perelomov, A.: Generalized Coherent States and Their Applications. Springer-Verlag, Berlin (1986)

[173] Peres, A. and Terno, D.R.: Quantum information and relativity theory. Rev. Mod. Phys. **76**, 93-124 (2004)

[174] Petersen, A.: The philosophy of Niels Bohr. In French, A.P. and P.I. Kennedy (eds.): Niels Bohr, A Centary Volume. Harvard University Press, Cambridge, MA (1985)

[175] Pitman, E.J. G.: The estimation of the location and scale parameters of a continuous population of any given form. Biometrika **30**, 391-421 (1939)

[176] Press, S. J.: Subjective and Objective Bayesian Statistics. Wiley (2003)

[177] Rice, J.A.: Mathematical Statistics and Data Analysis. Duxbury Press, Belmont, Cal. (1995)

[178] Robins, J.: A new approach to causal inference in mortality studies with sustained exposure period - Application to control of the healthy worker survivor effect. Math. Modelling **7**, 1393-1512 (1986)

[179] Robins, J.: Addendum to "A new approach to causal inference in mortality studies with sustained exposure period - Application to control of the healthy worker survivor effect." Comput. Math. Applic. **14**, 923-945 (1987)

[180] Rubin, D.B.: Bayesian inference for causal effects: The role of randomization. Ann. Statist. **6**, 34-58 (1978)

[181] Schumacher, B.: Quantum coding. Phys. Review A **51**, 2738-2747 (1995)

[182] Schweder, T., Hjort, N.L.: Frequentist Analogues of Priors and Posteriors. In: Stigum, B.P.: Econometrics and the Philosophy of Economics. Theory-Data Confrontations in Economics. Princeton University Press (2003)

[183] Searle, S.R.: Linear Models. Wiley, New York (1971)

[184] Selleri, F.: Quantum Paradoxes and Physical Reality. Kluwer Academic Publishers. Dordrecht (1990)

[185] Serre, J.-P.: Linear Representations of Finite Groups. Springer-Verlag, Berlin (1977)

[186] Shannon, C.E.: A mathematical theory of communication. Bell Syst. Tech. J. **27**, 379-423 and 623-656 (1948)

[187] Skjervheim, H.: Objectivism and the Study of Man. Universitetsforlaget, Oslo (1959)

[188] Smyth, G.K., Verbyla, A.P.: A conditional approach to residual maximum likelihood estimation in generalized linear models. J.R. Statist. Soc. B **58**, 565-572 (1996)

[189] Snedecor, G.W., Cochran, W.G.: Statistical Methods. 8th ed. Iowa State University Press, Ames (1989)

[190] Speed, T.: Restricted maximum likelihood (REML). In: Encyclopedia of Statistical Sciences. Update volume 1, Wiley, New York (1997)

[191] Stacey, R.D.: Managing Chaos: Dynamic Business Strategies in an Unpredictable World. Kogan Page, London (1992)

[192] Stein, C.: The admissibility of Pitman's estimator of a single location parameter. Ann. Math. Statist. **30**, 970-979 (1959)

[193] Stein, C.: Approximation of improper prior measures by prior probability measures. In: J. Neyman and L.M. LeCam (eds) Bernoulli, 1713; Bayes, 1763; Laplace, 1813, 217-240. Springer-Verlag, Berlin (1965)

[194] Stigler, S.M.: Do robust estimates work with real data? Ann. Statist. **5**, 1055-1098 (1977)

[195] Stone, M.: Right Haar measures of convergence in probability to invariant posterior distributions. Ann. Math. Statist. **36**, 440-453 (1965)

[196] Thomson. W.A., Jr.: The problem of negative estimates of variance components. Ann. Math. Statist. **33**, 273-289 (1962)

[197] Tjøstheim, D.: A communication relation for wide sense stationary processes. Siam J. Appl. Math. **30**, 115-122 (1976)

[198] Vapnik, V.: The Nature of Statistical Learning Theory. Springer, New York (1995)

[199] Varadarajan, V.S.: Geometry of Quantum Theory. Springer, New York (1985)

[200] Wahba, G.: Spline Models for Observational Data. SIAM, Philadelphia (1990)

[201] Weisberg, S.: Applied Linear Regression. Wiley, New York (1985)

[202] Wetterich, C.: Quantum entanglement and interference from classical statistics. http://xxx.lanl.gov/archive/quant-ph/0809.2671v1 (2008)

[203] Wetterich, C.: Probabilistic observables, conditional correlations, and quantum physics. http://xxx.lanl.gov/archive/quant-ph/00810.0985v1 (2008)

[204] Wetterich, C.: Emergence of quantum mechanics from classical statistics. http://xxx.lanl.gov/archive/quant-ph/0811,0927v1 (2008)

[205] Wigner, E.: On unitary representations of the inhomogeneous Lorentz group. Ann. Math. **40**, 149-204 (1939)

[206] Wijsman, R.A.: Invariant Measures on Groups and Their Use in Statistics. Lecture Notes - Monograph Series **14**, Institute of Mathematical Statistics, Hayward, California (1990)

[207] Williams, D.: Weighting the Odds. A Course in Probability and Statistics. Cambridge Univeristy Press, Cambridge (2001)

[208] Wolbarst, A.B.: Symmetry and Quantum Systems. Van Nostrand, New York (1977)

[209] Wotters, W.K.: The acquisition of information from quantum measurement. Ph.D. Thesis. University of Texas at Austin. (1980)

[210] Zacks, S.: The Theory of Statistical Inference. Wiley, N.Y. (1971)

Index